W9-AFC-051

ARTS AND TRADITIONS OF THE TABLE:
PERSPECTIVES ON CULINARY HISTORY

Albert Sonnenfeld, Series Editor

For a complete list of titles, see page 353.

THE TERROIR OF WHISKEY

ARTS AND TRADITIONS OF THE TABLE: PERSPECTIVES
ON CULINARY HISTORY

ROB ARNOLD

THE TERROIR OF WHISKEY

A Distiller's Journey Into the Flavor of Place

Columbia University Press / New York

Columbia University Press
Publishers Since 1893
New York Chichester, West Sussex
cup.columbia.edu

Library of Congress Cataloging-in-Publication Data
Names: Arnold, Robert, 1987– author.
Title: The terroir of whiskey : a distiller's journey into the flavor of place /
Rob Arnold.
Description: New York : Columbia University Press, [2021] | Series:
Arts and traditions of the table: perspectives on culinary history |
Includes bibliographical references and index.
Identifiers: LCCN 2020026158 (print) | LCCN 2020026159 (ebook) |
ISBN 9780231194587 (hardback) | ISBN 9780231550895 (ebook)
Subjects: LCSH: Whiskey—United States. | Distilleries—United States. |
Distillers—United States.
Classification: LCC TP605 .A767 2020 (print) | LCC TP605 (ebook) |
DDC 663/.50092—dc23
LC record available at https://lccn.loc.gov/2020026158
LC ebook record available at https://lccn.loc.gov/2020026159

Columbia University Press books are printed on permanent
and durable acid-free paper.

Printed in the United States of America

Cover design: Noah Arlow
Cover image: Kai Tilgner / Getty Images

What we know is a drop, what we don't know is an ocean.

—Isaac Newton

CONTENTS

THE TERROIR OF WHISKEY

INTRODUCTION

THIS IS A book about whiskey, a distilled spirit made from grain and (almost always) aged in oak barrels. Whiskey is essentially distilled beer, just as brandy is distilled wine. But specifically this book is about how *terroir*—a somewhat controversial concept with an unsettled definition—can influence the flavor of whiskey.

Terroir is a French word that describes how the flavor and characteristics of a crop (or livestock) are influenced by its environment. That includes the soil that layers the farm, the topography that shapes and molds its contours, and the climate in which it resides. At least, this is terroir in its most basic sense. And that is how I first thought of terroir when I began this book. Terroir, I gathered, was simply a romantic synonym for *environment* and how it influences the expression of a plant's genes.

All organisms have a genome, composed of thousands, millions, or billions of deoxyribonucleic acid (DNA) base pairs. DNA is the genetic code, the underlying blueprint for all traits, including the production of chemical compounds that we perceive as flavors. The environment can control how the genetic code is read, dictating whether certain traits are expressed. Grapes, grains, cheeses, dogs, or humans, we are all influenced by our environment—our terroir. If this sounds like science jargon, you may better recognize it as nothing more than the age-old debate of *nature vs. nurture.* Nature is what an organism is born with—its DNA. Nurture is how the reading and expression of the DNA is influenced by the environment.

The wine world is most famous for championing (and marketing) terroir. Napa Valley in California, Bordeaux in France, the Western Cape of South Africa: all of these regions possess their own terroirs—distinct soils, topographies, and climates—that influence the growth and flavor development of grapes. Because the concept of terroir is most firmly rooted in the wine industry, it's hard to discuss how the phenomenon applies to anything else without considering the parallels with wine. So while this book is about whiskey, it is necessarily also a book about wine.

* * *

I come from Louisville, Kentucky, a third-generation member of the whiskey industry. My grandfather and nearly every one of my uncles and great-uncles on my mother's side worked in bourbon. My grandfather was on the Brown-Forman company jet with the Brown family when they finalized the purchase of the Jack Daniel's distillery from the Motlow family in 1956. My great-great-grandfather was a brewmaster in Germany who brought his talents to Indiana at the end of the nineteenth century. Since 2011, I have been the master distiller at the Firestone & Robertson Distilling Co. in Fort Worth, Texas. Informally, we refer to our distillery as TX Whiskey. I joined the company as their first employee after opting out of a PhD program in biochemistry at the University of Texas Southwestern Medical Center.

Most parents would be distraught to hear their son was leaving medical research to make booze. But for me, it was a little different. When I called my mom to tell her I was leaving school to make bourbon in Texas she didn't say, "You can't leave school to make bourbon!" She said, "You can't make bourbon in Texas. You don't have the right water."

But Mom, it turns out you *can* make bourbon in Texas, and in any state for that matter. Kentucky has prime water for making bourbon, but so do many other states, including Texas. And Texas—with plenty of high-quality corn crops, hot summers, and winters that fluctuate between 30 degrees and 70 degrees Fahrenheit (often within the same day)—is just as suited to making bourbon as Kentucky.

We distilled our first batch of bourbon at TX Whiskey in February 2012. At the time, we were making about three barrels per batch, with a maximum capacity of three batches per day. At this scale, we were actually a relatively large *craft* distillery, but our output was a drop in the bucket compared to the big boys in Kentucky and Tennessee. I was personally just fine with making three to nine barrels of whiskey a day. But the distillery's original proprietors, Leonard Firestone and Troy Robertson, were not. They wanted to take the big boys head on, in quality *and* volume.

So in 2014 they bought a declining golf course and on its hundred acres we built one of the largest whiskey distilleries west of the Mississippi River. Our output is now forty barrels per batch, with a maximum capacity of three batches per day. In 2019, we sold TX Whiskey to Pernod Ricard, the second-largest wine and spirits group in the world. We are their largest whiskey distillery in the United States.

In 2016, I decided to combine whiskey and science into a second PhD attempt, this time at Texas A&M's distance plant-breeding program (*distance* meaning that I conduct my research at TX Whiskey and maintain my full-time position with the company while completing the degree), studying under the quantitative geneticist and corn breeder Dr. Seth Murray. My dissertation explores how genetic and environmental forces influence corn-derived flavors in whiskey. My hope is that the data generated can be used to breed and select new corn varieties specifically suited for whiskey production. Much of our research and the stories that surround it are covered in this book.

So through and through, I am a student, maker, and advocate of whiskey. Aside from when I am fortunate enough to tour a winery or vineyard—or when I steal a sip from my wife's wineglass—wine doesn't intrude on my mental or gustatory faculties. But over the years as I scoured the scientific literature on alcohol, visited wineries, and developed friendships with winemakers, I realized they pursued certain techniques that whiskey distillers were ignoring. Many of these pursuits are rooted in the concept of terroir and how the provenance of a place can translate to distinct flavors. These pursuits start not at the winery but in the vineyard. They start with the land and the grapes cultivated on it.

Imagine a trip to your local liquor store. If you're like me, you're taken aback by the sheer volume of wine bottles on the shelves. Hundreds or thousands of different wines from all over the world, labeled by region, grape variety, or both. How do you choose?

Maybe you select by grape variety, such as a merlot or a pinot grigio. Or perhaps you go by region, such as the terroirs of Sonoma Valley or Bordeaux. Or maybe you consider both, like a Napa Valley chardonnay. Your choices are immense, even though every bottle of wine starts with the same species of grape—*Vitis vinifera*. But this one species has thousands of expressions in the combination of varietal, terroir, and vintage (that is, the year on a wine bottle, which dictates the year the grapes were harvested).

On the shelf, there will be red wine made with the cabernet sauvignon grape variety grown in Napa Valley and the merlot grape variety grown in Tuscany. There will be white wine made with the sauvignon blanc grape variety grown in the Loire Valley of New Zealand and the chardonnay grape variety grown in the Adelaide Hills of Australia. There will be blended wines that contain grapes of different varieties and terroirs. And some wines—such as those from Bordeaux and Chablis in France—will be branded solely by the terroir, making the assumption that you will either know the grape varieties grown in that region, are curious enough to turn the bottle around and read the back-label description, or simply trust that these terroirs will provide enjoyable flavors regardless of which grape varieties were used. And even if you are just the average wine shopper and don't select wine based on a firm understanding of how flavor is dictated by particular varieties and terroirs, you will know enough—be it from experience, recommendation, or the bottle description—to decide whether you want a fruity and full-bodied cabernet sauvignon from Napa Valley, a chilled and fruity Italian moscato, or a spicy and floral French Bordeaux. The wine industry offers a seemingly innumerable number of choices based on varying combinations of grape varieties, terroirs, and vintages.

After choosing your wine, you then head to the whiskey aisle. You will see a few nationally recognized styles: Scottish, Canadian, Japanese, Irish, and American whiskies/whiskeys,[1] among others. And there are even distinct regional styles within some of these countries—Islay and Speyside whiskies from Scotland, Kentucky bourbon and Tennessee whiskey from the United States.

Some whiskeys are labeled by the type of grain *species*—barley for single malts, corn for bourbon—but you will find no mention of the grain *variety*. You might find some whiskeys labeled by where they come from, but almost without exception this has no bearing on where the grains were grown. When a wine is labeled Napa Valley, that is where the grapes were grown. When a whiskey is labeled Kentucky, the grains may have been grown as far away as Europe.

The truth is that historically, or at least in the last hundred years or so, whiskey distillers have not placed the same importance on the variety and terroir of their grains as winemakers have for their grapes. I don't mean to imply that whiskey distillers do not consider the quality of their grain. They do, but here "quality" means meeting certain specifications that align with the commodity grain trade: kernel density, percentage of damaged kernels, and percentage of foreign material (like dust and cobs). The consideration of flavor does not usually go beyond avoiding the bad ones—sour and must—and rarely to pursuing desirable and unique ones. So whiskey distillers typically grade their grain according to the quality metrics that the commodity grain trade follows. And this is because since the rise of the commodity grain trade in the early twentieth century, whiskey distillers have primarily sourced from it. The silos of the commodity grain trade hold an anonymous blend of varieties and terroirs from multiple farms, from dozens to hundreds to thousands. The distinctions of variety and terroir cannot survive the commodity grain silo.

* * *

At the inception of this book, I had a working definition of terroir. I treated it as this romanticized synonym for the environment (the soil, climate, and topography) in which something grew. And while the concept of terroir is rooted in wine, what does—or could—it mean for whiskey? But first I needed to answer a more direct question. Where did the word *terroir* come from?

Back when I was an undergraduate majoring in microbiology at the University of Tennessee in Knoxville, I would often visit my uncle and my grandfather in Bristol, a small city on the border of northeastern Tennessee and southwestern Virginia, best known as the home of the Bristol Motor Speedway. My uncle is a medical doctor, a student of Latin, an avid reader, and an amateur etymologist. I would bring along my homework, and he would help me navigate the etymology of the Latin names of microbes. These Latin names hold clues about the function of organisms. Consider the most common yeast used in the fermenting of wine, ale, and whiskey: *Saccharomyces cerevisiae*. "Saccaro" means sugar, "myces" means fungus, and "cerevisiae" means beer. String that together and you get "sugar fungus of beer." (You might also recognize a cognate in there for the Spanish word *cerveza*, which means "beer.")

I knew *terroir* was a French word and that French was a Romance language, a child of Latin. So, keeping in mind my uncle's lessons, I went to the dictionary. Now, if you are familiar with Latin, you may assume that *terroir* comes from the Latin word for land—*terra*. But it's not quite that simple. The French derivative of *terra* is in fact *terre*. In Spanish, it's *tierra*, and in Italian it is *terra*. So, *terroir*—while obviously related to the Latin word for land—is a bit more complex. It turns out that *terroir* stems from the Latin word *territorium*, which can roughly be translated as territory or an area of land with defined boundaries. While this was straightforward enough, it didn't really provide insight into how the word is regarded in the context of wine or any other food or beverage for that matter.

So I dug deeper, perusing English translations of French dictionaries and academic articles published by wine scientists and wine-marketing researchers. What I learned is that the word is contested—there is no satisfactory single-word translation or definition for it. It is uniquely French and seemingly impossible to translate into one English word. Even the French don't agree on what it means.

In French dictionaries you'll find multiple different definitions of *terroir*. As early as the seventeenth century, it did indeed simply refer to a territory or a region. But by the nineteenth century, its meaning grew, referring to a "small area of land being considered for its qualities or agricultural properties."[2] In his 1884 collection of essays *Les grotesques*, the French poet and dramatist Théophile Gautier described a hill with thin and rocky soil that produced excellent claret, a type of French rosé wine. He used the word *terroir* as a catchall for the entire set of characteristics of that hill.

The European Union established a legal definition in 2012. They claimed that products possessing terroir are those "whose quality or characteristics are essentially or exclusively due to a particular geographic environment with its inherent natural and human factors."[3] This evolved definition—for me—was more confusing than clarifying. So the word still described a physical piece of land and its soil, climate, and topography, but it also included "human factors." What did that even mean? It seemed so ambiguous, so amorphous, one of those words that makes sense to people who have used it all of their lives but could never explain it to someone who hasn't.

I decided terroir was not something I could learn solely from research papers or dictionaries. If I wanted to understand it, I would have to experience it, and I would have to taste it. I hope that—through this book—you are able do the same.

So here is what I propose. In this book, I will show how science can—to an extent—unravel the meaning and influences of terroir. You will join me on the journeys that I took—from the United States to Ireland and finally to Scotland—as I visited those farmers, winemakers, and distillers who are actively pursuing terroir. But ultimately you will only understand terroir—which is more than just flavor—by experiencing it. So I hope that my journey in this book can act as a guide—a blueprint—for how to design and embark on your own. I do hope that some of your journeys will take you on fantastic adventures around the globe, allowing you to meet the people who make the whiskey and then enjoy their art in the places that molded it. But a journey can also happen in the comfort of your own home. You can experience terroir simply by enjoying a dram of whiskey whose distillers have sought to capture terroir in some way. Or perhaps even multiple drams, as sampling a flight of whiskeys with diverging terroirs can be as enlightening as it is enjoyable. Regardless, I think you will find—as I did—that when a whiskey truly harbors a sense of place, then the experience it provides goes beyond just flavor. How exactly? Grab a glass of whiskey, and let's go find out.

PART I

FASHIONING FLAVOR, TASTING TERROIR

1
A FARM IN TEXAS

MY FIRST CONSCIOUS experience with terroir happened on a farm in the summer of 2014.

I was in Hillsboro, Texas, looking out at a cornfield. Large green blocks—the combines—crawled across the horizon. Everywhere I looked, a blue sky sat above a sea of dried cornstalks. Some stalks were taller than a man. Others were freshly cut and only inches off the ground. They covered every speck of visible land.

It must have been over 100 degrees Fahrenheit that day—harvest season for corn in North Texas does not fall in brisk autumn but during the hottest months of the year, usually in late July or early August. The sun was hot, and thousands of acres of cleared farmland meant I wouldn't find any shady trees for refuge.

I was visiting Sawyer Farms, which lies squarely between Dallas–Fort Worth and Waco. This particular farm supplies all of the barley, rye, wheat, and corn that we use to distill our various expressions at TX Whiskey, where I serve as master distiller. Established at the close of the nineteenth century, Sawyer Farms is now run by the fourth-generation farmer John Sawyer. For most of its existence, the farm grew grains for the commodity market.

A truck pulled up behind me and parked, and out came John Sawyer. Though successful in business, John never misses a chance to look the part of farmer. A relatively tall and slender middle-aged man with grayish-white hair, he unfailingly wears both Performance Fishing Gear shirts and hats that sport his farm's iconic logo, which reads "Sawyer Farms, 4th Generation." Driving his Ford F-350 up and down the embankments of his 3,500 acres in Hill County, he orchestrates his team of employees as they prepare, plant, cultivate, and harvest the grain crops.

I said hello, and together we watched the combine harvester cut back and forth across the fields, leaving a carpet of severed stalks rising no more than a few inches off the ground. From our vantage about fifty yards away, the only visible evidence that the machine was harvesting corn was the intermittent unloading of solid streams of yellow kernels from an attached swing auger into a grain hopper truck. For a farm of this size, the harvest would take over a week, weather permitting, and it would span many fields separated by creeks, roads, and even interstate highways.

"These combines, they actually contain GPS. It's like the GPS in your car, except in the combines we can integrate it with the steering mechanism itself," John said, with the appreciative tone of a man who had spent many years without such technology. "This isn't like the Siri GPS on your iPhone—it's not telling you to turn left when you reach the end of a row. Instead, it sets a course across the field and then maintains a straight line with an accuracy of just a few centimeters. When it reaches the end of a row, the driver works to pull the combine around, and then the GPS lines up the next row based on the previous cut. It's highly efficient and informative."

"Informative?" I asked. "Like it tells you where you have and haven't harvested?"

"Well, sure, it will tell you that. But maybe more importantly, it tells you exactly how many pounds of grain you're yielding—the yield for an entire field and for each individual section. This helps us target fertilizer and herbicide treatment regimens. If we know which sections in a field are prone to issues, we can target those specifically instead of treating the entire field. It saves me money, and it's more environmentally conscious. Less treatment means less excess chemical runoff."

I watched as the streams of yellow corn flowed from stalk to harvester to loading truck. I was here to see the corn I was buying to make bourbon. On this visit, in the early goings of our partnership with Sawyer Farms, we were still just a drop in the bucket of their total grain sales, the bulk of which was sold on the commodity market.

"So this yellow dent corn, at least the lot of it that we don't buy to make bourbon, will be used to make tortilla chips and grits?" I asked. I was trying to rattle off the few items I knew that used hard yellow dent corn instead of soft sweet corn.

"Oh, no," said John, "at least, I doubt it. This will all end up as animal feed, mostly for chickens and cows."

"Really? It doesn't get used as food? You can eat yellow dent corn, can't you?"

I turned corn into bourbon every day. Bourbon, by law, must contain at least 51 percent corn in the recipe, or what we call the *grain bill.* Legally, any class of corn can be used, from yellow sweet corn to blue flint corn. But yellow dent corn is the dominant type. To be honest, at this point, I didn't understand why yellow dent corn was the most prevalent type used to make bourbon. It was just the class of corn I was told to use by the master distillers who taught me. So given that whiskey made from yellow dent corn was indeed consumed by humans in bourbon, I figured the yellow dent corn not used for bourbon would naturally become human food.

"Sure, you can eat it," John said as we turned and walked back to his F-350. "But most yellow dent is sold on the commodity market to animal feedlots and ethanol producers. The corn grown on my farm is sold to the commodity grain market and used to feed the animals that humans eventually eat, not to feed humans directly."

"OK, well, I must be missing something. You provide us with food-grade corn. But if the majority of the yellow dent you're planting—and that basically all yellow dent corn farmers are planting—is grown to feed chickens and cows, then what separates *food-grade* yellow dent from *feed-grade* yellow dent?"

"Not much, really," John said. "Food grade will always be the densest kernels, but dense kernels can end up in feedlots as well. And food grade will be devoid of any mycotoxins, which come from fungal infections. But our corn is always devoid of mycotoxins. The main difference is just that food-grade grain has to be cleaned to remove the rocks, stalks, cobs, and insects. Feed-grade grain is, to a certain extent, allowed to have some of that stuff in there and therefore doesn't need to be run though a grain cleaner."

John explained that there were indeed some yellow dent corn varieties that were better suited for food than they are for feed. The suitability was mainly because certain varieties are easier to process, mill, and cook. And blue corn is primarily grown for tortilla chips because of its flavor and color. But many corn varieties can be either feed grade or food grade: it just depends on how well the farmers avoid foreign material and broken kernels during harvest and if they clean that stuff out after harvest. The same is basically true for all other grains.

"What about taste?" I asked.

"Well, other than the rocks and stalks, I guess the kernels themselves taste the same."

"But if one variety can be either grade, then the harvested lots that become food grade must have a better flavor. At what point is flavor considered?" I asked.

"I'm not sure if it ever really is. I thought that was your job." John threw his head back and laughed. "Us farmers don't get paid for flavor. Grain is a *commodity*. We get paid for yield. Maximizing how many bushels per acre can we make—that's the almighty goal."

* * *

Looking back on it, that experience at Sawyer Farms in 2014—when I heard that phrase "the almighty goal" and realized the relative apathy toward grain selection in whiskey production—was one of the most important days in my whiskey-making career. The questions it raised changed how I would select, cultivate, and store corn, wheat, rye, and barley at Sawyer Farms and the distillery.

John set me on a quest to understand the importance of place in whiskey—how the total environment in which a grain is grown changes its chemistry and, further down the line, influences the whiskey into which it is made. Winemakers call this *terroir.* This book is the catalogue of that quest. It would take me across the Atlantic, from the whiskey-making regions of America to Scotland and Ireland. But first it would send me to the birthplace of terroir: wine country.

I might be the first person ever to begin a whiskey journey in Northern California. But that's because the role of terroir in whiskey is hotly debated, even polarizing. Diving headfirst into the debates over terroir in whiskey would just lead me down a garden path of divisive and discordant opinions, many of which wouldn't be rooted in science. Terroir simply had no track record in whiskey. So I figured if I was going to explore terroir in whiskey, I first needed to explore it in wine.

* * *

There are many well-known winemaking regions in the world. I chose to visit the California counties of Napa and Sonoma.

What was my goal? To drink wine. But not just drink whatever I could find among the hundreds of wineries in Napa and Sonoma. No, I had a plan—a specific tasting experiment I wanted to conduct. But to do this, I needed the help of winemakers.

If I did this right, I hoped the experiment would give me concrete evidence that terroir does indeed influence flavor in wine, that it's not just a marketing buzzword. And maybe, based on what this evidence looked like, the experiment would also show me ways to detect or measure terroir in whiskey.

Now, I could have walked into any wine store and asked them to sell me a few different bottles that showed off expressions of terroir. But I reasoned that wouldn't be sufficient. The experiment wouldn't be sound. I needed assurance—beyond what a wine store clerk could provide—that the necessary control variables were present in my experiment. I had to ensure that terroir was the only independent variable in my tasting experiment, the only variable that would change from wine to wine.

My original plan was to investigate the notion of terroir in its simplest form: as an expression of how the environment—the climate, soil, and topography—changed grapes in the vineyard. Flavor production in grapes, I reasoned, could be influenced by the genetics of the grape, the terroir acting on a growing grape, the viticultural techniques used to grow the grapes, or some combination of all three. And at first, the wine terroir tasting experiment seemed simple enough.

2

THE PRODUCTION AND PERCEPTION OF FLAVOR

WINE IS CREATED through a process that is not—at least on the surface—that different from whiskey. Sure, there are aspects in which they diverge, sometimes drastically. But from a bird's-eye view, the creation of wine and whiskey share more similarities than differences.

The first step is to harvest and process a raw material. For wine, this raw material is grapes. For whiskey, it is grain.

Processing grapes means destemming and crushing them to create *must.* Red wine is made from whole must. White wine is pressed again to separate the *juice* from the skins and seeds.

Processing grain involves malting (sometimes), milling (always), and mashing (always).

Malting is when grain is allowed to partially germinate before being kilned and returned to a dormant state. Germination is when a grain kernel—which is technically a seed—begins to sprout. Kilning is the process whereby heat is applied to the germinating kernel, which removes moisture to the point that the kernel returns to a dormant state. The fuel for kilning is typically natural gas, but peat is also used sometimes, when the smoky flavors derived from it are desired in the whiskey. Unlike raw grain, malted grain is soft, brittle, sweet, and full of starch-degrading enzymes called amylases. When barley is used to make whiskey, it is more often than not malted barley. The primary exception would be Irish pot still whiskey, which uses a good amount of raw barley in the recipe. Corn, wheat, and rye are also common base grains for whiskey, depending on the specific style. Sometimes corn, wheat, and rye are malted, but more often they are used in their raw form.

Milling breaks apart the grain, creating a grist, which is basically coarse flour. The grist is combined with water in a process called *mashing*, and the result is a sweet liquid called *mash* or—if the mash is filtered—*wort*. By contrast, water is never added to the must or juice in winemaking—grapes already contain plenty of water.

After the raw material is harvested and processed, it is fermented. This is the process by which yeast and other microbes—namely, lactic acid bacteria—consume sugar and nutrients in the must, juice, mash, or wort and excrete alcohol, carbon dioxide, and flavor compounds as metabolic byproducts.

In wine and whiskey, the yeast species most commonly used in fermentation is called *Saccharomyces cerevisiae*. It has been our yeast of choice for making bread, wine, beer, and distilled spirits for thousands of years, even though we've only known of its existence (along with every other microorganism) since the seventeenth century (thanks to Antonie van Leeuwenhoek) and only confirmed the role it plays in fermentation since the nineteenth century (thanks to Louis Pasteur). So in essence, we did not choose *S. cerevisiae* to be the invisible actor of our bread and alcoholic drinks; the process of natural selection did, as *S. cerevisiae* is extremely proficient at both creating and sustaining alcoholic and acidic environments (bread would be alcoholic if the baking process didn't evaporate it away). Most other microbes are not very ethanol or acid tolerant and therefore die off in the environment created by *S. cerevisiae* fermentation. But there are exceptions, such as lactic acid bacteria, which can thrive in an alcoholic and acidic environment. Lactic acid bacteria are sometimes intentionally added in winemaking—kickstarting what winemakers call *malolactic fermentation*—but since they live naturally on grapes and grains, they will make their way into the process regardless. Malolactic fermentation is the process whereby lactic acid bacteria—namely, *Oenococcus oeni*, as well as multiple species of *Lactobacillus* and *Pediococcus*—convert tart-tasting malic acid into the more pleasing, softer lactic acid.

After fermentation, the making of wine and the making of whiskey diverge. Wine is allowed to settle so that dead yeast cells, skins, and seeds can fall to the bottom of the tank and be removed in a process called *fining*. In whiskey, the opposite is true. The fermented mash or wort, which is technically just beer at this point, is mixed up and transferred to a distillation system, which is aptly called a still. *Distillation* is the process of heating the mixture until the ethanol and flavor compounds evaporate and separate from the majority of the water and the grain or yeast solids that might be in the beer. The evaporated ethanol and flavor compounds travel up the still and to a condenser, where they return to a liquid state. This liquid is called *new-make* whiskey. Stills are often copper. Originally, this was because stainless steel did not yet exist and because copper is malleable and an effective heat conductor. But distillers also eventually realized that copper was important for flavor: the metal will react and bind with unpleasant sulfur-containing flavor compounds, successfully removing them from the distillate. The height, width, and overall shape of a still will also affect flavor.

From here, the manufacture of wine and whiskey converge again. New-make whiskey will, almost without exception, be transferred to oak barrels to age and mature. Wine isn't quite as invariable, but it too will often be aged in oak. However, there are indeed differences in how maturation proceeds between whiskey and wine.

Whiskey will usually spend years or even decades in barrels stored in a warehouse. While it depends on the location, whiskey-barrel warehouses—which are brick, wooden, or steel barns and not typically controlled with any type of heat or air conditioning—are often characterized by large seasonal temperature swings between hot and cold. These temperature fluctuations are actually desired, as they facilitate the expansion and contraction of the whiskey into the barrel's wood and then back out again, allowing the spirit to extract color- and flavor-inducing compounds from the oak.

Wine maturation is tamer, as the liquid might spend only a few months to a few years in oak, and usually in a stable environment, like a cellar or a cave. Wine is more prone to spoilage during maturation, as its ethanol concentration is not high enough to deter the growth of all microorganisms. Therefore, hot temperatures—which encourage microbial growth—are problematic.

Wine barrels are *toasted*. The cooper puts the barrel over a fire pit and controls the amount and length of heat per a specific recipe. In toasting, the barrel can be exposed to heat (via indirect flame) for as long as forty-five minutes, with temperatures ranging from 300 to 500 degrees Fahrenheit. The term *toasted* refers to the fact that the barrel never catches on fire.

Whiskey barrels are almost always (though not exclusively) *charred*. Charring is quicker and hotter, with heat exposure (via direct flame) lasting only thirty to sixty seconds and temperatures ranging from 500 to 600 degrees Fahrenheit.

Toasting and charring both break down the oak, but given that the temperatures and length of heat exposure are different, the concentrations and compositions of flavor compounds produced in each process will vary. According to Independent Stave Company, the largest barrel cooper in the world, toasting actually leads to a higher concentration of flavor compounds than charring. As such, many American distillers (including myself) are experimenting with oak barrels that are initially toasted and then very lightly charred (so as to stay compliant with the *new charred oak barrel* rule for the popular American whiskey styles, such as bourbon and rye whiskey).

The final step for wine and whiskey is bottling. Where many wines (at least those that are sealed with cork) can continue to age and mature in the bottle, whiskey does not. The main reason for this is that when a bottle is aged horizontally and in constant contact with the cork, oxygen will slowly diffuse into the bottle. As a result, oxidation reactions will slowly occur in the wine, which can affect the concentration of nearly all flavor compounds (aside from certain fusel alcohols). It should be noted that while brief aging for a few months is desired for almost all wines, certain styles (including many white wines) do not stand up well to prolonged aging and will quickly deteriorate as oxygenation proceeds.

Many whiskeys are sealed with screw tops instead of cork. Therefore, oxygen cannot enter the bottle, and oxidation reactions will not occur. So, assuming the bottle has sat unopened for a prolonged period, the whiskey will taste nearly the same—even after months, years, or decades—as it did the day it was bottled. For those whiskeys that are sealed with cork, they are typically stored upright versus horizontally, meaning that the whiskey is not in contact with the cork. Therefore, oxygen will not readily diffuse into the bottle. However, for whiskeys sealed with cork, as cork will over time dry and shrink, oxygen will eventually enter, but this occurs over longer spans of time: years or decades. But whether cork or screw top, if enough oxygen does enter the bottle, then the whiskey will begin to change from the same oxidation reactions that take place in an aging bottle of wine.

* * *

There are basically three overarching styles of wine—red, white, and pink (or rosé). The color variations have to do with both the variety of grape and whether the entirety of the grape—the skins, seeds, pulp, and juice, collectively referred to as *must*—is fermented. This is the case with red wine, and pigments in the skin—the most notable being phenolic compounds, such as flavonols, anthocyanins, and tannins—impart the color. In white wine, the grapes are pressed to separate the juice, which is fermented alone. Pink wine incorporates some color from grape skins, but not so much that it qualifies as red wine. Whiskey has no counterpart process to this. All whiskey is as crystal clear as water after distillation. The color comes from aging in the barrel, specifically from the wood's tannins and caramelized wood sugars.

Both wine and whiskey styles are usually specified and marketed by their regional origin. The most general way to describe a wine from France is to call it French wine. This is almost always specified down to the actual region where the grapes were grown. French wine made in the Bordeaux region is classified as Bordeaux wine. Wine made in California is called California wine and often—if the grapes all come from the same region in California—narrowed down to the region, such as Napa Valley, or even subregions within a region, like St. Helena or Yountville. Many of these wine regions and subregions—including Bordeaux, Napa Valley, St. Helena, and Yountville—are geographically bounded by legislation. The French certification is known as "appellation d'origine contrôlée" (AOC), while the U.S. certification is known as American Viticultural Area (AVA).

Just the same, whiskey is often named after the country or region where it's made. Scotch whisky is whisky made in Scotland. Irish whiskey is made in Ireland. Canadian whisky must be made in Canada. Some styles can only be produced in a certain country or region. For example, bourbon is a style of whiskey that must be produced in the United States. Contrary to common misconceptions, bourbon can be produced outside of Kentucky, its birthplace. In fact, it can be produced anywhere in the United States, and it's been that way since the birth of the American whiskey industry centuries ago. Further, Congress didn't even declare bourbon as a whiskey style distinct to the United States—granting it geographical protection—until 1964. During Prohibition, from 1920 to 1933, some distillers moved south of the border and produced Mexican bourbon.

Whiskey doesn't have a system that exactly mirrors wine's AOCs or AVAs, but there are regions within countries that further specify a style from the general style of the country. Within Scotland, for example, there are five distinct whisky-making regions: Speyside, Highland, Lowland, Campbeltown, and Islay. While there are plenty of exceptions to the rule, the five regions are considered to produce styles with distinct flavors. Speyside is known for producing fruity, floral, and delicate whiskies, whereas Islay is famous for its heavily peated, smoky, and medicinal flavors. American whiskeys are almost always further specified geographically based on the state. Kentucky whiskey must be produced in Kentucky and aged at least one year there. Tennessee whiskey must be produced in Tennessee and by Tennessee state law must use the Lincoln County charcoal mellowing process made famous by Jack Daniels.

In the whiskey industry, what the word *produced* on the label means is often debated. It can sometimes mean nothing more than where it was bottled. In the United States, among craft whiskeys, this is often a point of tension as sometimes a state designation is used even though the only production process that took place inside that state was bottling. But recently, there has been a push for regulations to mandate that whiskeys with a state designation must be distilled and aged in that state. New York and Missouri have taken this a step further, requiring a majority of locally grown grains to be used for two of their state-specific styles—Empire rye and Missouri bourbon, respectively.

Beyond geography, wines and whiskeys are usually specified by the varietal of grapes or species of grains from which they are fermented and distilled. Almost all wine is produced from the same species of grape—*Vitis vinifera*. But within that species are thousands of varieties. The most common are called the International Varieties or the Noble Grapes, and these are the household names: merlot, pinot noir, chardonnay, sauvignon blanc, riesling, and cabernet sauvignon. When a wine is produced using an overwhelming majority of one of these, it is usually labeled by the varietal. A merlot wine is made from merlot grapes. And many wines are labeled by both their geography and their grape, like the famous Napa Valley chardonnay.

Interestingly, many Old World wines, like those from Bordeaux, will not be labeled according to the varietal. This is tradition, and it's also because laws usually dictate which grape varieties can be grown within different regions. Bordeaux wine can only be produced from certain varieties, usually cabernet sauvignon, cabernet franc, merlot, petit verdot, and malbec. Bordeaux wine producers—and their avid drinkers—believe listing the grape varieties on the label would be redundant.

Whiskey can be produced from *any* species of grain. The four most common are corn (*Zea mays*), barley (*Hordeum vulgare*), rye (*Secale cereal*), and wheat (*Triticum aestivum*). There are ongoing projects and low-volume releases of whiskeys made from other grain species such as milo (*Sorghum bicolor*, which is synonymous with grain sorghum) and triticale (*Triticale* spp., which is a hybrid of wheat and rye). But unless you've specifically sought out a milo or triticale whiskey, every whiskey you've ever enjoyed is some combination of corn, barley, rye, and wheat.

Malt whiskey, the most well-known of which is scotch single malt whisky, must be produced from malted barley. In scotch single malt, the word *malt* is legally defined to mean that the whiskey is produced from 100 percent malted barley. The word *single* means that it is the product of a single distillery. A scotch blended malt whisky still contains 100 percent malted barley, but it is a blend of whiskies produced at more than one distillery. Given the prominence of scotch single malt, most other countries follow their legally defined guidelines, always with a grain bill of 100 percent malted barley. Irish malt whiskey and Japanese malt whisky, which have gained popularity over the past decade, are also produced from 100 percent malted barley.

The Irish also have a distinct style of whiskey that, like bourbon, may only be produced in Ireland: pot still whiskey. The name is somewhat confusing, as both Irish malt whiskey and Irish pot still whiskey are produced in pot stills. What separates the two styles is that pot still whiskey must contain raw, unmalted barley along with some malted barley.

In Canada, whiskies are usually mashed from 100 percent of the same grain. Canadian rye whisky is 100 percent rye, for example. But the Canadian industry has a long, storied, and impressive track record of blending, so most bottles of Canadian whisky are blends of multiple grain types. But the blending happens in the final stages, after the whisky is harvested from the barrel, and not in the beginning stages during mashing, as is commonly done in American whiskey.

American whiskey regulations are unique in that the classification styles require only 51 percent or more of a single grain type. Whereas Canadian rye whisky is 100 percent rye, American rye whiskey can be as little as 51 percent rye. American wheat whiskey must be 51 percent or more wheat, and bourbon whiskey must be at least 51 percent corn. One exception is American corn whiskey, which must be made from at least 80 percent corn. You may have noticed that under these rules a grain bill over 80 percent corn could become either bourbon or corn whiskey. What separates them is the barrel. Bourbon must be aged in new charred oak barrels, as do American rye and wheat whiskeys. Corn whiskey, conversely, must be unaged,

aged in used barrels (e.g., barrels previously used to age bourbon or sherry), or aged in new toasted barrels.

The strict barrel requirements of American whiskey are unique, as no other country mandates the use of new charred barrels for making any style of whiskey. As a result, most used bourbon barrels are sent to Scotland, Ireland, Canada, and Japan, where they will age those country's whiskeys, often on more than one occasion. These countries will also use barrels that first aged sherry or other fortified wines, but the percentages pale in comparison to former bourbon barrels, called *ex-bourbon* in the trade.

* * *

That's a lot of information. The main takeaway I want you to have is this: the flavor compounds of wine come from the grape, the fermentation byproducts, and—if barrel aging is employed—the oak barrel. This is true regardless of the style of wine. While bottle aging (or even aging in stainless-steel tanks before bottling) will manipulate and change flavor compounds—through chemical reactions—it does not introduce any new compounds.

In the same way, the flavor compounds in whiskey come from the grain, fermentation byproducts, and the oak barrel. While water can technically introduce flavor compounds (think of bad tap water that tastes like dirt and earth, flavors that come from compounds like geosmin and 2-methylisoborneaol), more often than not water is filtered through activated carbon to remove any organic compounds and off-flavors. And similar to bottle aging in wine, while distillation will manipulate and change flavor compounds—through chemical reactions—the process does not actually introduce any new compounds.

We also perceive the flavor of wine and whiskey in the same way. Our sense of flavor is created through the interplay of multiple sensory modalities as they detect an external stimulus. Every sensory system can influence our perception of flavor—even sound! Think of how the pop of a cork anticipates a certain flavor expectation. That said, for flavors, smell and taste are the most influential of the senses.

The bulk of a flavor experience is created when odorant receptors in the nose and taste receptors on the tongue detect chemicals called flavor compounds. These compounds bind to the receptors and activate them, leading to a chain reaction of chemical signals that terminate as a signal in the brain. When multiple odorant and taste receptors are activated—as is the case when food and drink is consumed—multiple signals sent to the brain combine to create the sensation and realization of flavor. We detect flavor the same way we hear a musical chord. When we hear a G chord on the piano, we don't hear the individual G, B, and D notes. We hear the chord. While sensory experts can break a flavor down into its constituent parts, our physiology is tuned to recognize flavor cohesively. To most of us, Coca-Cola has the flavor of, well, Coca-Cola. Those trained in sensory methods, however, can hone in on and separately detect the lemon, vanilla, and cinnamon flavors that, among others, make up the flavor of Coca-Cola.

Which matters more to our perception of flavor: smell or taste? It turns out that it is actually smell. Taste is relatively limited in what it can perceive: only sweet, sour, salty, bitter, and umami. Taste is even more limited with regard to whiskey because the high ethanol content desensitizes the taste receptors. Scientists debate the range of smells humans can detect, but they estimate we can identify anywhere from ten thousand to more than one trillion different aromas.[1] A huge variance, but in any case we can smell far more than the five basic tastes.

This might not come as a surprise: most people have noticed that they can't taste food when they have a cold or when they plug their nostrils. When your nose is blocked, the flavor impact of aroma loses its punch. You might still taste the sweetness of an apple, but you will not taste the dozens of flavor compounds in the apple that truly create the experience of *apple*.

Odorants (also called volatiles) are those flavor compounds in wine and whiskey (and any other food or drink) that are volatile, therefore allowing them to reach the odorant receptors in the nose. This process happens through both orthonasal olfaction (when odorants enter your nasal cavity through sniffing) and retronasal olfaction (when odorants enter your nasal cavity from inside the mouth through breathing out).

There are hundreds of potential flavor compounds in wine and whiskey, many of which are odorants that belong to diverse chemical classes: organic acids, esters, ketones, aldehydes, terpenes, pyrazines, acetals, alcohols, lactones, and sulfur compounds. Although many of these odorants are present only at very low concentrations—from a thousandth to a billionth of a gram per liter of whiskey—their sensory thresholds are often correspondingly low. A sensory threshold is the concentration of a substance necessary for you to smell or taste it. So even a low concentration of an odorant (either in combination with similar odorants or individually) can significantly change a food or drink's flavor.[2]

* * *

If my wine terroir tasting was going to reveal any information that could translate to whiskey, it would be vital to understand the flavor chemistries of both drinks. However, most scientific research related to alcohol up to now has focused on wine. There's a lot of reasons for this—financial funding, culture, and the integrated nature of certain wine industries with their local universities, like the University of California–Davis with the California wine industry and the University of Bordeaux with the French wine industry. In addition, wine carries a certain prestige, whereas whiskey is often seen as the poster child of alcoholism. This is, of course, not true—wine is just as likely to be abused as whiskey. But it's been easier for the ivory tower of academia to convince their donors that wine research is warranted and just.

After wine, beer is the next most studied. Whiskey is somewhere far behind them both. It's true that Seagram's did conduct and publish significant whiskey-science-related research in the first part of the twentieth century. And the Scotch Whisky Research Institute does actively conduct research and publish its findings. But the volume of scientific literature on whiskey is—compared to wine and even beer—sparse.

However, because of the shared or similar aspects, techniques, and processes of wine and whiskey (and beer, for that matter) production, I reasoned that many of the same flavor compounds, principles, and scientific findings from wine could potentially apply to whiskey, too. If I could discover which grape-derived flavor compounds are changed by terroir in wine, would that chart some type of chemical roadmap for terroir's effect on flavor in whiskey? I wasn't sure. But the approach seemed logical enough.

I first saw this idea of a chemical roadmap in a 2017 paper published in *Science* titled "A Chemical Genetic Roadmap to Improved Tomato Flavor."[3] Led by the principal investigator and University of Florida horticulturist and plant breeder Dr. Harry Klee, the paper revealed which compounds—and the genes that control their production—are responsible for the flavor of tomatoes. Perhaps I could do something similar for whiskey, by way of wine.

Perhaps your first reaction to my idea is that it doesn't make much sense. While whiskey and wine both use the same yeast for fermentation and similar barrels for aging, grains and grapes are nothing alike. How could starting with grapes provide any kind of roadmap for grain?

Well, going by look, feel, and even taste, grains and grapes seem totally different. But genetically, they are much closer than you might think. A grain kernel is technically the fruit of a cereal grass. When you eat fried tortilla chips or Cheetos, you are eating (at least in the botanical sense) fruit.

Structurally, the grape berry and the grain kernel are remarkably similar. Fleshy fruits like grapes have pericarps that are soft and palatable. Pericarps are the ripened ovary of the fruit. Animals eat the pericarp and dispose of the seed. Dry fruits—such as grain kernels of barley, wheat, rye, and corn—have pericarps that harden and fuse to their seeds, and so animals eat both the pericarp and the seed of a dry fruit.

Many of the compounds in grapes and grains influence flavor indirectly, as they will be metabolized by yeasts and serve as precursors to fermentation byproducts. Conversely, certain small molecules produced by grapes and grains can potentially impart flavor directly. They will find their way into the must, juice, mash, or wort, and subsequent fermentation, distillation, and aging will not alter their structure to any great extent. Additionally, certain chemical precursors in grains will undergo chemical reactions such as oxidation, the Maillard reaction, Strecker degradation, and caramelization during high-temperature malting, mashing, and distillation. These chemical events can produce flavor compounds with fruity, fatty, roasted, nutty, caramel, and malty flavors. The important point, though, is that the pathways and processes for how grape- and grain-derived flavor compounds make their way into a wine or whiskey are actually much more similar than they are different.

This reality provides an opportunity for a comparative study. Scientists routinely compare the genes and traits of unknown or understudied organisms to those of known and exhaustively studied organisms. We often study gene functions in yeasts to understand what genes in our own bodies do. For example, both yeasts and humans share much of the same metabolic machinery for turning sugar into adenosine triphosphate (ATP), which is the molecular currency of energy. Indeed, from an evolutionary standpoint, we are actually not such distant relatives of yeast. Our genes are close enough that computational methods can often determine which ones are of the same descent. Such genes are said to be *homologous* to one another. Because we can't experiment on humans, poking and probing them in controlled environments, changing one variable at a time to see how the presence and expression of certain genes are affected, we use yeasts. Yeasts don't complain or file lawsuits, and they reproduce rapidly every few hours or so.

Given that grains and grapes are both fruits and share some common ancestor, their flavor chemistries should share some homology. And if they do share some flavor compounds, and if studies have shown that those shared compounds are influenced by terroir in wine, then it follows that terroir might also influence them in whiskey.

At least, this was my hypothesis. And my plan was to produce a chemical roadmap drawn from homology, from common genetic descent, and from the ample scientific literature on the terroir of wine.

* * *

As a distiller, I knew enough about the flavor chemistry of whiskey. But as the ideas for this book began to form, months before I visited northern California wine country, I realized that if I ever hoped to develop this roadmap, I would have to brush up on my whiskey chemistry and would need a crash course in the chemistry of wine flavor.

Based on what I would discover, I would have to categorize flavor compounds by their source—grapes and grains, fermentation, and oak. This way I could place the flavor compounds into buckets and highlight the ones that were most promising for changing the flavor of whiskey. I would focus on the bucket of flavor compounds that were derived from grapes. Of those compounds, I could determine which ones were reportedly changed by terroir and then pick out the ones (or their close derivatives) that are also present in whiskey.

From there, I would taste and smell whiskeys from around the world, looking for flavor correlations between what was reported in wine, what I personally experienced in wine, what was reported—as rare as that might be—in whiskey, and what flavors I discovered in those whiskeys. Where possible, a chemical analysis could give me concrete proof of what my taste and smell receptors were telling me.

Of course, it wouldn't be that simple. Science never progresses as planned. But at a minimum, I had a hypothesis—and the makings of a scientific investigation.

3

THE CHEMISTRY OF FLAVOR

THE FIRST STEP on my journey was to learn about the major classes of flavor compounds in wine and where they originate, what influences their presence and concentration, and how they overlap with those found in whiskey. I started by scouring the scientific literature—reading research reports and reviews—to provide some foundation for my chemical roadmap idea.

Unless you're a chemist, chances are that reading a simple regurgitation of what I'd read will be extremely dry. And when analyzing a wine or a whiskey, even professionals do not start with chemistry. We start with flavor.

So instead of a literature review, this chapter will be a virtual wine tasting with me (or, ideally, a real one), a journey through the flavors and chemistry of grapes.

We are going to pour four glasses of wine for our tasting. While you are free to choose your own brands, try to choose one muscat (dry or sweet), one sauvignon blanc, one unoaked syrah, and one oaked cabernet sauvignon. We will break the tasting up into two sessions, each of which has a different goal for experiencing flavor.

Our first session is straightforward. It won't require you to think or analyze much. Take the first glass, swirl it around a bit, bring it up to your nose, and take a sniff. After you've experienced the aroma, go ahead and take a sip. Let the wine coat your mouth, and then move it around some, which will aid in volatilizing the flavor compounds to activate retronasal olfaction.

Now repeat these steps for the other glasses.

What do you smell and taste? If you are an experienced wine drinker, you might already be trying to break down the wines, picking out nuances unique to each. If you are one of these people, take a step back and clear your head of any flavors you were starting to describe. If you are not one of these people, and you haven't tried to describe the nuances of the wine, then that's perfect! What I want you to do is *not* pick out any nuances or differences. Instead, I want you to describe what the wines smell and taste like in the most basic terms.

What do you taste? I know what my experience is—*they all smell and taste like wine*. And for this session, that's a perfect reaction! Regardless of varietal, vineyard, yeast, barrel, vintage, bottle aging, or anything else that might influence flavor, all wines have the basic flavor impression: *wine*. The wine scientists refer to this as the "global odor"[1] of wine—the aroma shared by all wines. They describe it as *vinous*, which can further be defined as "slightly sweet, pungent, alcoholic, and a little bit fruity."[2]

While most drinkers might assume that the flavor of wine derives mostly from grapes, in 2008, a wine researcher named Vicente Ferriera showed that this global odor of wine is actually a combination of eighteen compounds that come almost solely from fermentation—primarily, a mixture of higher alcohols (also called fusel oils), esters, fatty acids, diacetyl, and acetaldehyde (table 3.1).[3] Only one compound derived from grapes directly contributes to this global odor—β-damascenone (*cooked apple*). The alcohols, esters, fatty acids, diacetyl, and acetaldehyde produced during fermentation are present at relatively high levels, measuring in at parts per million (milligrams per liter). β-damascenone is once again the outlier, in that it is typically present only at concentrations of one to two parts per billion (micrograms per liter) in red wine and five to ten parts per billion in white wine.

So where do these global odor compounds come from in wine, what are their individual flavors, what affects their production and concentration, and do they play any potential role in the flavor of whiskey?

TABLE 3.1 THE EIGHTEEN FLAVOR COMPOUNDS RESPONSIBLE FOR THE GLOBAL ODOR OF WINE

COMPOUND CLASS	FLAVOR COMPOUND	SOURCE	AROMA
Norisoprenoid terpene	β-damascenone	Grape	Cooked apple
Fusel (higher) alcohol	Isoamyl alcohol	Fermentation	Banana
Fusel (higher) alcohol	Phenethyl alcohol	Fermentation	Floral, rose
Fusel (higher) alcohol	Methionol	Fermentation	Meaty
Ester	Ethyl isovalerate	Fermentation	Apple, pineapple
Ester	Ethyl 2-methylbutanoate	Fermentation	Apple
Ester	Ethyl isobutyrate	Fermentation	Citrus, strawberry
Ester	Ethyl butyrate	Fermentation	Pineapple, mango
Ester	Ethyl acetate	Fermentation	Fruity, ethereal
Ester	Ethyl octanoate	Fermentation	Floral, banana, pineapple
Ester	Ethyl hexanoate	Fermentation	Fruity, apple
Ester	Isoamyl acetate	Fermentation	Banana
Organic (fatty) acid	Butyric acid	Fermentation	Rancid
Organic (fatty) acid	Octanoic acid	Fermentation	Fatty
Organic (fatty) acid	Hexanoic acid	Fermentation	Cheesy
Organic (fatty) acid	Isovaleric acid	Fermentation	Cheesy
Aldehyde	Acetaldehyde	Fermentation	Green apple
Ketone	Diacetyl	Fermentation	Buttery

All of the global odor flavor compounds aside from one are produced by yeast during fermentation. Yeasts do not mean to deliver the intoxicating effects and delicious flavors that their human curators are after. What we call alcohol and flavor compounds are simply waste products to a yeast cell—they are what's left and expelled after the microbe has metabolized nutrients into as much energy and usefulness as possible. (There is some evidence that yeasts produce ethanol to kill any competing microbes near it, but that is still more theory than fact. In fact, if ethanol production by yeast does serve an evolutionary purpose, it may be linked to the fact that yeast can use ethanol for energy when sugar is depleted.) The nutrients metabolized by yeasts for energy and growth are the same ones humans use for the same purposes—sugars, fatty acids, and amino acids. All fermentation-derived flavor compounds—whether they are in the global odor group or not—are basally derived from the metabolism of sugars, fatty acids, and amino acids.

Ethanol is the most prominent alcohol in any alcoholic beverage, but its flavor is relatively bland, contributing a mild, *solvent-like sweetness*. However, ethanol influences the intensity and perception of other flavor compounds. For example, it suppresses fruity aromas from esters and enhances spicy flavors from volatile phenols.[4] And of course, ethanol is the alcohol responsible for the feeling of intoxication. But ultimately, ethanol does little to contribute directly to the flavor of wine or whiskey.

As opposed to ethanol, higher alcohols—so called because they contain more carbon atoms than ethanol—possess a range of flavors. The three included in the global odor group of wine are isoamyl alcohol (*banana*), phenethyl alcohol (*floral*, *rose*), and methionol (*meaty*).

There are eight esters in the global odor group. Seven of them are ethyl esters, meaning they are formed from ethanol and an acid. The one outlier is isoamyl acetate, which belongs to the class known as acetate esters, which form from acetic acid (vinegar) and an alcohol. Esters are almost universally *fruity*, *floral*, and *waxy*. While esters are indeed synthesized by grapes and grains, they are usually present at such low concentrations that their effect on the flavor of a wine or whiskey is insignificant.[5] All of the important esters in wine and whiskey are produced by yeasts or formed through the chemistry of alcohols and acids at some point in the process, such as distillation (in the case of whiskey) and during maturation.

There are four fatty acids in the global odor group, and they possess a range of flavors that are not exactly inviting: butyric acid (*rancid*), octanoic acid (*fatty*), hexanoic acid (*cheesy*), and isovaleric acid (*cheesy*).

The three remaining compounds in the global odor group are all from three distinct classes.

Acetaldehyde is the lone aldehyde, and it is usually described as having a *green apple* flavor.

Diacetyl is a vicinal diketone, and it possesses a distinct *buttery* flavor.

Last, β-damascenone—which smells of *cooked apple*—is the only nor-isoprenoid terpene in the global odor group, and it is also the only one that does not derive from fermentation. It derives from grapes, specifically from carotenoid molecules in the berry.

After reading these past few paragraphs, you may be wondering: If each of these global odor compounds has its own distinct (and sometimes undesirable) aroma, then why do they in combination result in the general (and desirable) *vinous* flavor present in all wines? The answer is nothing more than the old adage that the whole is greater than the sum of its parts. Individually, they do indeed possess their own flavors. But collectively, they act in a concerted, synergistic manner, and the result is *vinous*. They "are no longer perceived as single entities, because their aromas are fully integrated to form the complex concept of wine aroma."[6]

So now that we know where these global odor compounds derive from, the next step is to understand what can influence their concentration in wine.

One of the first gas chromatography studies in wine was monumental for showing that different yeast strains produced different levels and types of higher alcohols.[7] While it had been known through practice and sensory analysis studies that different yeast strains created different flavors, this early gas chromatography work revealed some of the specific compounds responsible for the variation. Later studies have found that the concentration in wine of every compound in the global odor group is at least partly dependent on yeast strain, fermentation temperature, and nutrient composition.

That the nutrient composition of a must, juice, mash, or wort has influence over the production of fermentation-derived flavor compounds is an important point in the context of terroir. What nutrients a grape contributes to a must or juice, or that a grain contributes to a mash or wort, is dependent on the characteristics of the grape and grain. Those characteristics will be dictated by variety and growing environment (be it soil, climate, agronomic techniques, etc.). What this means is that terroir can potentially influence the production of flavor compounds produced by yeast during fermentation.

For example, while all grapes and grains contain proteins, the concentrations and compositions will vary among varieties and species. During fermentation, yeast can convert amino acids—the building blocks of proteins—into higher alcohols and esters through a set of metabolic processes known as the Ehrlich pathway.[8] So, even though the resulting higher alcohols and esters may not derive directly from grapes or grains, they do derive from their amino acids. Therefore, it can be reasoned that grapes and grains with higher concentrations of proteins (which can be degraded into their amino acid constituents during malting and mashing) and amino acids may end up producing wines and whiskeys with more ester and higher alcohol flavors compared to grapes and grains with lower levels of proteins and amino acids. Further, the specific composition of the proteins, and therefore the amino acids, will also lead to varying flavors. For example, the amino acid leucine will be metabolized into isoamyl alcohol (*banana*) and ethyl isovalerate (*apple*, *strawberry*), while the amino acid valine will be metabolized into isobutyl alcohol (*vinous*) and ethyl isobutyrate (*citrus*, *strawberry*).

It's not just the makeup of the proteins and amino acids in the fermentation that can affect the production of flavor compounds. Even the specific makeup of sugars will change the flavor. In wine, glucose and fructose sugars dominate the must or juice. In beer, maltose (a disaccharide composed of two glucose molecules) dominates, but it is complemented by glucose and maltotriose (a trisaccharide composed of three glucose molecules). Research has shown that there are no significant differences in ester and higher alcohol concentrations between glucose and fructose metabolization. However, higher alcohol concentration disproportionately increases when the ratio of sucrose increases, and conversely, higher alcohol and ester concentrations decrease as maltose ratios increase.[9] Why different sugar metabolisms lead to varying levels of higher alcohols and esters is still being researched.

β-damascenone is a norisoprenoid terpene reported to be produced from the degradation of neoxanthin, which is a carotenoid.[10] Present in plants and algae, carotenoids are color-producing pigment compounds that absorb light energy for photosynthesis and protect chlorophyll from photo-damage. The composition and concentration of carotenoids in grapes is reported to be influenced by grape variety, soil characteristics, climate, and viticultural practices.[11] Hello, terroir. I know you're there somewhere.

* * *

While it's useful to understand the origin of these global odor compounds, as well as what can affect their concentration—such as terroir—there is a catch. Ferriera and his team actually found that the mixture of global odor compounds has a certain innate "buffer" to it, meaning that the flavor does not change (or changes very little) if one compound is present at very low concentrations or dosed back in at exaggerated levels. So in reality, the concentration of these global odor compounds doesn't seem to influence the general wine flavor of *vinous*. Considering this point, the main takeaway for me was that while the global odor compounds are crucial to creating the flavor of wine (and maybe whiskey), they will always be present in wine—regardless of any contribution from terroir—and the concentrations of each can vary widely while still collectively delivering the same *vinous* flavor. Therefore it seemed reasonable that terroir's role in controlling the compounds' production—if it had any—wouldn't be important for creating flavor differences among wines.

There are, however, two potential exceptions—isoamyl acetate and β-damascenone. Ferriera found that when the former was omitted, the global odor compound mixture experienced a noticeable decrease in fruity flavor. When the latter was omitted, the mixture experienced a noticeable decrease in overall flavor intensity. So while the other sixteen compounds could probably be ruled out as being responsible for any flavor differences among

TABLE 3.2 IMPORTANT IMPACT COMPOUNDS AND THEIR POTENTIAL SOURCE(S) IN WINE AND WHISKEY

COMPOUND CLASS	SOURCE
Acetal	Fermentation
Aldehyde	Grain/grape, fermentation
Ester	Fermentation
Fusel (higher) alcohol	Fermentation
Ketone	Grape/grain, fermentation
Lactone	Grape/grain, fermentation, oak
Pyrazine	Grape/grain
Organic (fatty) acid	Grape/grain, fermentation
Terpene	Grape/grain, fermentation
Norisoprenoid terpene	Grape/grain, fermentation
Sulfide	Grape/grain, fermentation
Phenol	Grape/grain, fermentation, oak
Furan	Grain, oak
Tannin	Grape, oak
Pyrone	Oak

wines—whether they are influenced by terroir or not—isoamyl acetate and β-damascenone were still in the running. Indeed, they are perhaps more than just in the running. They might be prime targets, as both isoamyl acetate and β-damascenone are also reported to be important contributors to flavor in many styles of whiskey.[12] (Side note: Given that β-damascenone is present at concentrations that are at least one-thousandth of the other global odor compounds, it highlights that a compound's concentration does not always dictate its importance in driving flavor—the sensitivity of the human nose to the compound is equally important.)

Further, there are many other flavor compounds produced using the same metabolic pathways of grape growth and yeast fermentation—those same pathways that are potentially influenced by terroir—that are indeed responsible for the flavor differences that exist among all wines. These other flavor compounds are called *impact compounds*. Let's focus on them by starting our second tasting session. Table 3.2 summarizes some of the most important classes of impact compounds and their potential source(s) in wine and whiskey.

* * *

Now for the second tasting. As with the first session, smell and taste the wines again. But this time, we'll move beyond the *vinous* flavor present in all of them. Each one of these wines smells and tastes different. Why?

It's because *impact compounds* in each of the wines are present at high enough concentrations that they reveal flavors beyond the global odor. These impact compounds can act individually or synergistically, but either way, they are responsible for the flavor variations that exist among wines. And, just maybe, among whiskeys.

So now that we've established that our four wines possess different flavors (no shocker there), let's dive a little deeper into your glass of muscat. If you need a refill, now would be a good time for that.

Smelling and tasting the muscat, do you detect a distinct *floral* aroma? That aroma is almost certainly from a compound called linalool, which is present at relatively high levels in muscat.

Linalool belongs to a class of compounds called terpenes, which are a large and diverse group of highly odorous compounds produced by plants either to deter herbivores or to attract the predators and parasites of herbivores. While there are other terpenes in wine beyond linalool, they all generally impart *floral* and *citrus* aromas. Terpenes and their precursors originate chiefly in the grape skins, and as the grape matures on the vine, their concentrations increase. During fermentation, terpenes are freed from the grape and make their way into the wine matrix. Terpene composition is often determined by the grape variety, and while most grapes contain some of the fifty or so terpenes that have been identified, they are especially prevalent in muscat wines and rieslings. Linalool specifically is also a critical contributor to the characteristic flavor of wines from Galicia, in northwestern Spain. And beyond linalool, rose-*cis* oxide is another monoterpene that imparts a sweet, floral, green aroma to Gewürztraminer wine. Cool climates and shaded vineyards usually result in decreased terpene production. Beyond the genetics of the grape variety or the environment of the vineyard, certain viticultural techniques such as shoot trimming and leaf trimming (both of which increase light penetration to the fruit zone) can maximize terpene production.[13] While linalool and rose-*cis* oxide are not to date reported in whiskey, other terpenes—and specifically the norisoprenoids, such as β-damascenone—appear to be very important contributors to flavor in whiskey.

Let's move on to the sauvignon blanc. Pick up your glass of sauvignon blanc, give it a swirl, assess the aroma, and take a sip. It will certainly have flavors like the muscat (regardless of whether you chose a sweet or dry one), and *floral* may be the shared flavor that comes to your mind. And this is no surprise. Many white wines possess a *floral* aroma, and as in muscat specifically, linalool may be responsible for this. But unlike in muscat, linalool is not alone. Many white wines smell and taste *floral* because of the presence and synergistic effect of multiple impact compounds, including (but not limited to) linalool, γ-lactones, ethyl cinnamates, α-ionone, β-ionone, and α-damascone. So while your muscat and sauvignon blanc may have similar *floral* aromas, they won't be identical.

Some of the γ-lactones are produced by yeast, which will metabolize fatty acid precursors that are themselves produced by the bacteria that coinhabit the must or juice fermentation. Other γ-lactones are produced by the chemical oxidation of fatty acids at some point in the winemaking process. γ-lactones do indeed appear to be important contributors of flavor in bourbon, rye whiskey, and malt whiskey.[14]

Ethyl cinnamates are esters of ethanol and cinnamic acid, the latter of which is an important structural component of grape berries that protects against ultraviolet radiation. Given the origin of cinnamic acid, it's likely that this ester forms through reactive chemistry in the wine itself during or after fermentation and not inside of the yeast cell. Ethyl cinnamates have indeed been identified as important flavor compounds in at least two styles of whiskeys—bourbon and rye.[15]

Like β-damascenone, α-ionone, β-ionone, and α-damascone are norisoprenoids (which are actually a subclass of terpenes). All norisoprenoid terpenes are derived from carotenoids. When these compounds are in the grape and at the start of fermentation, they are bound to sugars and therefore not volatile—that is, they will not contribute to aroma. However, during fermentation and as the wine ages in an oak barrel, a stainless-steel tank, or even in the bottle, the norisoprenoids are released from the sugar molecules, becoming very important contributors to a wine's flavor. Similar to other terpenes, the environment of the vineyard is important in determining the production levels of norisoprenoids in grapes, with warmer climates and increased sunlight leading to higher concentrations. Once again, many whiskey styles also appear to contain norisoprenoids.[16]

Another sip of the sauvignon blanc. What other flavors can you detect? Do you perhaps pick up on some flavors of *green mango*, *tropical fruit*, and *box tree* (which can also be described, rather unfortunately, as *cat pee*)? If so, these are likely caused by a group of sulfur-containing compounds called thiols (specifically, 3-mercaptohexan-1-ol, 3-mercaptohexylacetate, 4-methyl-4-mercaptopentan-2-one, and 4-mercapto-4-methylpentan-2-one). These compounds are produced by yeasts from amino acid metabolism, specifically the amino acid cysteine, which contains a sulfur atom. While these four thiols are not reported as important contributors to flavor in whiskey, other sulfur-containing compounds are, such as dimethyl sulfide (*cooked corn*), dimethyl disulfide (*vegetables*), and dimethyl trisulfide (*meaty*). These three are formed from similar amino acid metabolisms in yeast as the thiols. And remember, the concentration and composition of amino acids is ultimately dictated by the characteristics of the grape or grain.

One last sip. Do you detect some *green*, *vegetal*, *herbaceous*, or *bell pepper* aromas? If so, these are likely produced by the group of compounds called methoxypyrazines. Specifically, isobutyl-methoxypyrazine (IBMP) is often considered to be the most important of the group, followed by isopropyl-methoxypyrazine (IPMP) and sec-butyl-methoxypyrazine (SBMP). Beyond sauvignon blanc, these methoxypyrazines also impart their distinct flavors to cabernet sauvignon and certain wines from Bordeaux. While the details surrounding their biosynthesis in grapes are still being researched, they do originate in the fruit, and it appears these varieties have a greater propensity to produce them than others. Consequently, only nanograms per liter concentrations are found in wine. But even at these minuscule levels, they still can have a meaningful effect on flavor. They also appear to be produced in grains, and while their detection in whiskey is sparse, one report has shown that IPMP is a potentially important odorant in bourbon.[17]

* * *

While white wines are delicious, they are typically served chilled, and I assume at this point that your glasses of muscat and sauvignon blanc have warmed up some. So let's move away from them and turn to the reds.

Just as in white wine, impact compounds in red wine are responsible for all the flavor variations that exist. However, unlike in white wine, individual compounds are usually not responsible for distinct nuances in red wine. Instead, these nuances are the result of large groups of impact compounds acting in a synergistic manner.[18] Some of the most important of these are the volatile phenols.

Pick up your unoaked shiraz. Spend some time assessing its flavors. Do you notice that while it is still certainly fruity, it's not quite as fruity as the white wines? Further, do you detect some distinct *spicy*, *clove*, and *smoky* flavors? While there is an important terpene called rotundone that can contribute some of these notes, it's likely that a group of compounds known as the volatile phenols are largely responsible. Volatile phenols can suppress *fruity* flavors while simultaneously imparting their own distinct flavors.

Given that this shiraz has not been aged in oak, the only source of volatile phenols would come from the metabolism of hydroxycinnamic acids by yeast. Hydroxycinnamic acids are components of lignin and lignan, and they are primarily found in the skin and flesh of grapes. The most abundant form in grapes is caffeic acid, followed by coumaric acid, ferulic acid, and caftaric acid. While hydroxycinnamic acids themselves possess flavor, they are not present in high enough concentrations in grapes for us to taste them. But when they are metabolized by yeasts, their byproducts contribute greatly to the *spicy*, *clove*, and *smoky* flavors of certain wines, such as those found in your shiraz as well as many other red wines. Hydroxycinnamic acids are found in grains, too, and reports show that different barley varieties contain different concentrations of hydroxycinnamic acids, which could translate into different levels of volatile phenols in whiskey.[19]

Finally, let's move to our last wine, the cabernet sauvignon aged in an oak barrel. The influence of oak on the flavor of a wine can be drastic, and in fact, the only impact compound in red wine that is said to deliver its own distinct aroma, instead of working with many others in a synergistic manner, is derived from oak. It's called *cis*-oak lactone. As you smell and sip the oaked cabernet sauvignon, you will most likely detect sweet notes of *vanilla*, *oak*, *burnt sugar*, and *coconut*. You have the oak barrel to thank for these flavors.

Of all the ingredients and processes in wine and whiskey, the oak barrel may be the most iconic and recognizable. Nearly every style of whiskey is aged in oak, which is responsible for all of the whiskey's color and much of its flavor. While not all wine is aged in oak barrels, it is a very common practice that contributes distinct and desirable flavors to those that are. The flavor compounds from oak are derived—for both wine and whiskey—from degradation of the wood's celluloses and hemicelluloses (wood sugars), tannins, lignins, and lipids. This degradation is facilitated by two factors: microbial breakdown from the months- or years-long outdoor seasoning of oak staves and through the toasting or charring of the oak barrel with indirect heat or direct ignition. While cellulose is another important component of oak wood—as well as the most abundant natural polymer on Earth and critical for the wood's structural strength—it is of such a strong and crystalline nature that it is largely resistant to microbial and thermal hydrolysis. Historically, it's been reported that cellulose is neither degraded nor extracted into wine or whiskey, and as such it has little influence on flavor.[20] However, a recent report from 2018 showed that cellulose does indeed break down during the charring process and ultimately affect the flavor of whiskey.[21]

Hemicellulose, unlike cellulose, is amorphous, relatively weak, and therefore prone to hydrolysis. Seasoning, toasting, and charring break down hemicellulose into a variety of caramelization products, and these provide *sweet*, *caramel*, *maple*, and *burnt sugar* flavors (from compounds such furfural, 5-methylfurfural, cyclotene, and maltol).

Like cellulose, lignin is also a widely abundant natural polymer that supports the wood's structure and water conduction thanks to its ability to link with—and fill the spaces between—cellulose, hemicellulose, and pectin components. It is constructed from phenolic building blocks, namely, guaiacyl and syringyl. Upon seasoning, toasting, and charring, guaiacyl breaks down into *sweet*, *vanilla* flavors (from compounds such as vanillin, vanillic acid, and coniferaldehyde), and syringyl degrades to *smoky*, *spicy* flavors (from compounds such as guaiacol, syringic acid, syringaldehyde, and sinapaldehyde).

Small amounts of lipids—oils, fats, and waxes—are the source of the very important compounds previously discussed called oak lactones (also called whiskey lactone, as it was first isolated and identified from whiskey). Lipid degradation produces two stereoisomers of oak lactone—*cis*-oak lactone and *trans*-oak lactone. Stereoisomers are pairs of molecules that have the same number and types of atoms (so the same chemical formula), but their atoms are arranged differently in three-dimensional space. It's this different arrangement that leads to their variations in flavors and intensity. While both will deliver flavors of *coconut*, *vanilla*, and *oak*, *cis*-oak lactone is more intense and possesses *rose-like* aromas, whereas *trans*-oak lactone contains nuances of *celery*. Collectively, oak lactones are major contributors of oak-derived flavors in wine and whiskey.

The hydrolysable tannins (gallotannins and ellagitannins) are important contributors to mouthfeel in both wine and whiskey. Tannins are also important in whiskey for promoting oxidation, leading to the formation of *floral* acetals.

The composition and concentration of oak-derived flavor compounds varies by oak species, forest location, seasoning location and length, and degree of toasting. Based on the oak species and forest location, the composition and concentration of hemicellulose, lignin, and lipids will vary. The location and length of seasoning and the degree of toasting will have different effects on degradation and the ultimate flavor compounds that result.

* * *

This tasting has walked us through the commonalities of wine, the compounds responsible for the flavor variations among certain wines, and which wine flavor compounds might overlap with whiskey. But this isn't the full story.

For example, there are other acids not yet discussed that are potentially very important for flavor in wine. Wine fermentation produces a myriad of volatile organic acids (called volatile acids), but different types of nonvolatile organic acids (called fixed acids) are also produced by grapes. Fixed acids are predominately tartaric, malic, and citric acids. While all three are present in grapes, citric acid is present only at trace levels. Succinic acid, uniquely among the fixed acids, is produced by yeast during fermentation. Fixed acids add sourness—think of lemons, which are rich in citric acid, or pineapples, which are rich in succinic acid. Winemakers also use lactic acid bacteria in a process called malolactic fermentation to convert the tart malic acid into the softer-tasting lactic acid. While these fixed acids are important for taste in wine, they do not evaporate during the distillation of beer into whiskey, and therefore they don't appear at all in whiskey.

While the oak barrel can contribute tannins, so can the grape. Tannins fall into two categories—hydrolysable and condensed. Hydrolysable tannins primarily come from the oak barrels. Condensed tannins come largely from grape seeds, pericarp flesh, and skins. While all tannins deliver a drying, bitter sensation, some are smoother and less coarse than others. That said, similar to fixed acids, while grape-derived tannins are important in wine, grain-derived tannins do not evaporate during distillation and therefore do not play a role in whiskey.

Moving beyond wine, there are of course some processes in whiskey that have no equal in wine. Malting is one such process.

While grain can be used in its raw form, it can also be malted. By definition, malt is germinated grain that has been dried before it can grow into a new plant. The main goal of malting is to turn hard, starchy grain into a form that is softer, sweeter, and full of starch-degrading enzymes called amylases. The malting process proceeds in three steps: steeping, germinating, and kilning. In steeping, harvested, dried, and dormant grain is exposed to multiple water immersions over the course of a few days to increase its moisture content and bring the grain back to an active state. After steeping, the high-moisture grain moves to a germination vessel. Over the course of four days or so, the grain (which is working under the assumption that it needs to become a new plant) begins to grow. The critical amylase enzymes are produced, as it begins to tap into its own starch reserves for food. The final stage is kilning, where the maltster halts germination by applying hot air, bringing the temperature of the malt to anywhere from 100 to 220 degrees Fahrenheit, which in turn dehydrates it. This desiccation reverts the grain to a dormant state. The newly produced enzymes are still present, but they too go dormant.

The malting process creates flavors that can end up in whiskey. The extent and length of heat applied during kilning can create vastly different malt classes with varying degrees of flavor and color. Many of the flavor compounds produced during the malting process are caused by Maillard reactions. While you may not have heard of these reactions by name, they are universally recognizable and responsible for the browning appearance and roasted flavors of many household foods, for example, bread crust, coffee beans, grilled meats, and French fries. Chemically, Maillard reactions occur between sugars and amino acids under high-temperature conditions. The products of Maillard reactions are incredibly diverse: hundreds have been identified. Furfural is an important one that imparts grainy and almond-like flavors. What's key, though, is that the type of amino acid that takes part in the initial reaction is largely responsible for the flavor compounds ultimately produced. Therefore, species and varieties with varying levels and concentrations of amino acids may produce different flavors even under identical malting conditions. α-dicarbonyl intermediates of the Maillard reaction will also cause Strecker degradations of amino acids to certain aldehydes during kilning. Further, if the kiln temperatures are high enough (230F or higher, which occurs during a high-temperature kilning technique known as "roasting"), sugar caramelization will occur. Roasting of malt is not common for spirits but is typical for the raw materials used in stout production.

Further, the fuel for the heat can impart flavor to the malt. While flavorless hot air generated by coal- or gas-fired furnaces is commonly used, some distilleries still choose to use smoky, hot air partly derived from a traditional fuel source, such as peat. When most plants die, they are decomposed entirely by microbes. However, in some environments—such as bogs that are waterlogged, acidic, and overly anaerobic—decomposition does not go to completion. The coastal regions of Scotland and Ireland often produce such conditions. Thousands of years ago, when plants in these bogs died, their decomposition was only partial, and the result we have today is peat. It's similar to coal, except there are still plenty of aromatic chemicals locked inside, and these are the source of the flavor compounds that are imparted by peat. The flavors of peat vary widely, and the nuances depend on both the type of vegetation that once inhabited the bog and the degree of decomposition.

The use of peat as a fuel source during the kilning process results in drastically different flavors than the use of natural gas. Peat is full of volatile flavor compounds that possess peaty, phenolic, meaty, smoky, burnt, and medicinal characteristics imparted by a wide range of cresols, guaiacols, and other volatile phenols. These flavor compounds will become adsorbed on to the malt (meaning that they form a film on the malt's surface) during the kilning process. The exact flavors imparted will depend on a number of factors. One will be the degree of thermal degradation of the peat when it is burned during kilning. The conditions of the kiln are also important—the longer the grain is exposed to peat smoke, the more intense the infusion of the peat's flavors.

In Scotland, studies have been conducted to assess how different peat sources affect flavor.[22] Peat from four locations—Islay, Orkney, St. Fergus, and Tomintoul—was analyzed using gas chromatography mass spectrometry. The results showed that the flavor compounds varied based on location. Where Tomintoul had relatively high levels of *medicinal* and *antiseptic* flavors (from phenols), St. Fergus contained a relatively high abundance of *spicy*, *sweet*, and *smoky* flavors (from guaiacols and syringols). A specific type of guaiacol called acetovanillone, with the flavor of *vanilla*, was prominent in the peat from Islay and Orkney.

Another process in whiskey making that has no equal in wine making is mashing. During mashing, whether with malted or unmalted grains, the high temperatures will (similar to high-temperature kilning) result in the production of aldehydes, largely from lipid oxidations and Strecker degradations. Aldehydes are important grain-derived flavor compounds in whiskey that arise from lipid oxidation. They can impart a wide range of flavors to whiskey, such as 2-methylbutanal (*chocolate*), 2,4-decadienal (*soapy*), 2-nonenal (*cardboard*), 2,6-nonadienal (*cucumber*), 2,4-decadienal (*fatty*), and nonanal (*soapy*).[23] The latter three have also been reported as important flavor compounds in some wines.[24]

* * *

There are other important flavor compounds in wine and whiskey, but what is covered here introduces the bulk of the most important classes and their origins. But even when we pinpoint the chemistry through advanced techniques, and even when those data are married to sensory analysis, it's not fully possible to explain all of the flavors in a glass of wine or whiskey . . . or beer or coffee or tea or anything else. For now, there is still some mystery to the origins of flavor. The important point is that flavor compounds (or their precursors) in wine and whiskey come from grapes or grains, fermentation byproducts, and maturation in oak barrels. And any time a certain thread of the wine or whiskey tapestry is pulled on—whether that thread is linked to the ingredients, the process, or both—the entire flavor picture can shift. Flavor is as sensitive as it is complex.

But it is possible to bucket flavor compounds based on their source: grapes and grains, fermentation, or oak. And it is possible to link *flavors* to specific compounds or groups of compounds. And most encouraging, it appears that many of the flavor compounds that derive from grapes and are reportedly changed by terroir are also present in grains and whiskeys.

My next step was to explore these shared flavor compounds. And I decided to start by experiencing terroir in wine firsthand, to prove to myself that it was real. I wanted to taste its existence and experience its tangibility. If I wasn't convinced that it was legitimate or something that could be scientifically investigated, then I would drop the project. But if I could taste its existence in wine, then that would be enough evidence for me to explore it in whiskey.

4

THE WINE TERROIR TASTING

AT FIRST, BEFORE heading to the Northern California wine country, I thought I would be testing the idea that terroir was an expression of the environment, specifically, an expression of the climate, soil, and topography in which the grape was grown. But as I read through the academic literature, I realized it was not that simple, at least not always. Wine places so much importance on grape variety, and more often than not, it is impossible to separate the varieties from the terroirs in which they thrive. Terroir appeared to be just as much about how particular varieties thrive and grow in specific environments as it was about the environment itself. It wasn't nurture alone but the intertwining of nurture and nature.

Think of it like this. Let's say we have two varieties of corn—Reid's Yellow Dent and Bloody Butcher. Let's say we plant the two varieties on the same farm in Kentucky. They each grow into a plant and produce healthy ears. We harvest and shuck the ears, and then we taste the kernels. Lo and behold, the two varieties do, in fact, taste different. We'll say that Reid's Yellow Dent produced the flavor profile *sweet corn*, and that Bloody Butcher produced the flavor profile *fruity*. The flavor profiles are an expression of certain genes responsible for producing certain compounds. Some of these flavor genes might not be influenced by environmental forces, but at least a portion are.

Now, let's say we take those same two corn varieties and plant them on a different farm, one in Texas. They grow, we harvest, we shuck, and we taste. But this time Reid's Yellow Dent and Bloody Butcher both produce the flavor profile *fruity*. The environment of this second farm in Texas influenced the Bloody Butcher variety in the same way the first farm did. But Reid's Yellow Dent responded to this second environment differently, and its flavor profile shifted.

Now, finally, let's say we take the same two varieties and plant them on a third farm, one in New York. Once more, they grow, and we harvest, shuck, and taste. This time Reid's Yellow Dent does not taste like it did when grown on the other farms. On this third farm in New York, it produces the flavor profile *sulfury*. And the Bloody Butcher also behaves differently, and now it has produced the flavor profile *floral*.

So we have four different flavor profiles—*sweet corn*, *fruity*, *sulfury*, and *floral*—produced by two varieties grown across three environments. If terroir was simply *nurture*—that is, if the environment is solely responsible for the flavor profile—then we would expect to have six different flavor profiles.

But we have only four (see table 4.1).

This is the thought experiment I conducted when the meaning of terroir really began to widen for me. Terroir is not just the environment: at a minimum, it is how different crop species and varieties express flavor *through* their environment. The diversity of grain and grape varieties, I realized, was a crucial character in the terroir story.

Quickly after I realized that the grape or grain variety was just as much a part of the concept of terroir as the environment, I began to realize just how fast the term *terroir* could widen to include, well, seemingly every aspect of the vineyard. Climate, soil, topography, agricultural practices. The neighboring plants, animals, and microbes that also call the vineyard home. The species and varieties of grape. And the elusive human element. The entirety of the vineyard and farm system was potentially a player. Terroir wasn't just an expression of the environment. It was the flavor of an ecosystem.

This new approach to terroir was exciting, but would it simply add a level of complexity that would be impossible to understand? Maybe. But there

TABLE 4.1 HOW NATURE (GENETICS) AND NURTURE (ENVIRONMENT) COLLECTIVELY IMPACT FLAVOR

	FARM		
VARIETY	Kentucky	Texas	New York
Reid's Yellow Dent	Sweet corn	Fruity	Sulfury
Bloody Butcher	Fruity	Fruity	Floral

was the study from Vicente Ferreira's group showing that all wine does indeed have a conserved *vinous* aroma ("slightly sweet, pungent, alcoholic and a little bit fruity")[1] that comes from eighteen flavor compounds—the global odor compounds.[2] Of these eighteen, only two are reported to impart flavor nuances when their presence and concentration are altered—β-damascenone (*cooked apple*) and isoamyl acetate (*banana*). Beyond the global odor compounds, there are impact compounds, which are responsible for delivering characteristic aromas to specific wines. Some of these impact compounds are universally present in the majority of wines, and they act in synergy—versus individually—to produce flavor. The concentration variations among these universal impact compounds are responsible for some of the flavor nuances among wines. Other impact compounds are so influential that they can individually deliver unique flavor notes. These are often found in only a few styles of wines (sometimes only in one). Examples of these highly influential impact compounds would be linalool, which delivers a *flowery* aroma to muscat, and 3-mercaptohexan-1-ol, which imparts a *green mango* (or *cat pee*) aroma to sauvignon blanc, cabernet sauvignon, and merlot.

So, I reasoned that any nuances from wine to wine were attributable to variations in the presence and concentration of β-damascenone, isoamyl acetate, and/or any of the impact compounds. If I could understand how terroir influenced the production of these flavor compounds, then maybe I could begin to unravel the burgeoning mystery of terroir.

* * *

I had this idea of a chemical roadmap—guided by homology—to the terroir of whiskey. I had a good grasp on the flavor chemistry of wine and whiskey. I knew that the genetics of a crop and the environment in which it grows—collectively, terroir—will affect its flavor. I was now ready for the wine terroir tasting.

But ready—and thirsty—as I was, if I didn't set up the tasting correctly, I could easily misidentify flavor differences as coming from terroir when in reality they came from another "nonterroir" variable, like winemaking techniques, yeast strain, or the oak barrel. Or I might easily lose the ability to discern which aspect of terroir (environment, variety, vineyard management) was responsible for any flavor differences. If I wanted to learn how terroir changed the flavors in wine that come just from the grape, then I would need to set up a strict experiment, one that controlled for ingredient and process variables across a sufficient number of wine bottles.

THE EXPERIMENT

My plan was to use three different grape varieties grown on three different vineyards and assess for flavor variation. I would try to assign broad flavor profile rankings to each wine in an attempt to simplify the results. The design would look like a tic-tac-toe box. Each cell would pair grape variety and vineyard. Three varieties, three vineyards: nine wines (see table 4.2).

First I had to identify the variables for which I would have to control.

The first I identified was rooted in the definition of grape species: a variety versus a clone. Different varieties (think merlot and pinot noir) possess different flavors, but so do different clones of the same variety. Merlot clone ISV-V-F-2 tastes different than merlot clone ISV-V-F-6.

All grape varieties start as a single plant grown from a single seed. A seed is created through sexual propagation, when the male flower from one

TABLE 4.2 EXPLORING WINE TERROIR WITH VARIETY AND VINEYARD REPLICATIONS

Variety 1—Vineyard 1	Variety 2—Vineyard 1	Variety 3—Vineyard 1
Variety 1—Vineyard 2	Variety 2—Vineyard 2	Variety 3—Vineyard 2
Variety 1—Vineyard 3	Variety 2—Vineyard 3	Variety 3—Vineyard 3

grapevine parent fertilizes the female flower of another. Sexual propagation mixes DNA from the mother grape and the father grape. When this seed is planted, the grapevine that grows will be genetically distinct from its two parents: a new variety. Every grape seed is technically a new grape variety. If you were to plant the seed of a pinot noir grape, then whatever grows will no longer be pinot noir—it will be something new. The seed might have half of its pinot noir's mother DNA, but the other half would come from the father, which might be an entirely different variety or species.

Even if the two parents were both pinot noir, the genetic recombination of plant sex would produce seeds that would not yield pinot noir grapevines. This pinot-noir–by–pinot-noir cross might be closer to pinot noir than if the cross were between pinot noir and merlot, but it would still be different enough genetically to warrant its own name. The first plant for each of the many grape varieties we know today was selected for its desirable characteristics, like hardiness or delicious flavor. After proving their worth, these varieties warranted a spot in the vineyard, at the bar, and at the table. Some obtained exalted status, like the noble grapes: sauvignon blanc, riesling, chardonnay, pinot noir, cabernet sauvignon, and merlot.

Once a desirable grape variety has been identified, it is isolated and bred through asexual vegetative propagation. Asexual propagation means that each offspring of the grape will contain only the single parent's genetic information. It will be a clone.

Winemakers use one of three techniques to clone the grape. One way is to cut and replant the *shoot*, which is basically the stem, from a mother vine. A new grapevine will grow from it. This process usually sees the baby grapevine clone develop in a nursery before it's replanted in the vineyard to mature and bear fruit. A faster route is to remove the canopy (which is where the grape bunches and leaves emerge) and much of the trunk from one grapevine and *graft* it with the cutting of a new grapevine. Grafting is the process by which two plant parts are physically affixed until their tissues fuse and they become a single plant. Grafting a cutting will bear fruit quicker than planting a cutting. The last technique is called *layering*: bending the shoot of a grapevine into the ground and burying it in the dirt. The buried vine will grow into an independent plant.

But sexual propagation is not the only way to introduce the unique characteristics of hardiness and flavor that come from genetic variation. While vegetative propagation avoids the genetic reshuffling of sexual propagation, it does not avoid the random DNA mutations that occur when the plant grows. Imagine a vineyard with thousands of merlot vines that all started off genetically identical. Grapevines are not discarded after harvesting their fruit—they stay in place to bear fruit again the next year. Over time, the grapevines accumulate mutations in their DNA. This happens to all organisms. Most human cells are actively growing and multiplying, and they can accumulate DNA mutations. Solar radiation from the sun is one very common cause of DNA mutations in human cells. It's no different with grapevines. Their cells are actively growing, and solar radiation, replication errors, and mutagenic compounds produced by neighboring microbes, insects, or pesticides can damage their DNA. Sometimes the cells do their job correctly and repair the DNA. But sometimes they don't, and a mutated DNA sequence is born. Mutations are often harmless or insignificant, and their influence on flavor is imperceptible. Other times the mutations can decrease hardiness or deliciousness—and when these mutations occur, the grapevine is culled from the vineyard.

But every so often, the mutation brings new and desirable characteristics. Maybe an improvement in flavor or a tolerance to a pest or disease. These mutated vines are cut and propagated, and though the grapevine will still be merlot (in our example) it will be a unique merlot: a clone with a set of characteristics slightly different from its neighboring vines. But every merlot clone is still a merlot—they are all still the same variety. Mutations almost never accumulate to a point where a new variety arises. These mutations are more often deleterious than advantageous, so well before they have accumulated enough to classify the vine as a new varietal, it would lose its ability to function normally and die.

There are some notable exceptions. Pinot gris and pinot blanc, for instance, are two varieties that arose from DNA mutations in pinot noir vines. These mutations changed the flavor and color such that the viticulturist who propagated these clones hundreds of years ago deemed them separate varieties. But from a geneticist's point of view, these three pinots are all merely different clones of the same varietal.

What all this meant for my tasting was that it wouldn't be enough to know the variety of grape—I would need to know the clone within the variety.

Another variable for which I needed to control was vintage. The year noted on a wine label is the year the grapes were harvested from the vineyard, not the year the wine was barreled or decanted. Seasonal fluctuations from year to year mean that the flavors of one vintage differ from the next. Merlot grapes grown on a vineyard in Napa Valley in 2015 would taste different—to some extent—than the grapes that would grow from the same vine in 2016. In some respects, I saw vintage as a feature of terroir and potentially a variable that would be better to investigate than to control for. However, wine—unlike whiskey—continues to age and mature in the bottle. A 2015 Napa Valley merlot might taste different than a 2016 Sonoma County merlot, and this difference could indeed be from the environmental variations of 2015 and 2016. But it also could be from the one-year difference in bottle aging. Given that, I decided to control for the vintage variable whenever I could.

Then I had to control for variables outside of the grapes themselves. Production techniques can change a wine's flavor. Did the winemaker use a commercial strain of yeast or leave the vats open for wild yeast and bacteria to ferment the wine? If, for example, a Napa Valley merlot was fermented with the widely used Lalvin EC-1118 (Prise de Mousse) yeast strain, while a Sonoma County merlot was fermented naturally with whatever yeast and bacteria were living on the grape skins and in the winery—then they would taste different, and it could obscure the flavor effects of terroir. Even if the Sonoma County merlot was fermented with a commercial yeast strain, if it wasn't Lalvin EC-1118, then the experiment to explore terroir would still be compromised.

And what about the oak barrels? If the Napa merlot was aged in oak but the Sonoma merlot was not, then I might be tasting only the mellowing and maturing of the barrel, not terroir. And I couldn't ignore the *species* of oak, either. If the Napa merlot was aged in French oak (*Quercus petraea* or *Quercus robur*) and the Sonoma merlot was aged in American oak (*Quercus alba*), flavor differences could be largely attributable to the varying oak species, not the terroir of the grapes. For example, French oak is known to deliver spicy and tannic flavors, whereas American oak provides sweet and coconut flavors.

As if that wasn't enough, my oak barrel controls had to go even further: "French oak," which many winemakers use, wasn't specific enough. There are six main French forests from which a wine barrel's oak lumber can come: Limousin, Vosges, Nevers, Betranges, Allier, and Tronçais. Based on which forest the wood is harvested from, a barrel could be built with *Q. petraea*, *Q. robur*, or both. And since this is an experiment in how terroir affects vegetation, well, I reasoned that flavors in the oak staves could also be influenced by not just the species but the environments of the forest. So not only would I have to control for the oak species that made the barrel, I would also need to control for the forest from which that oak was logged.

This was getting out of hand. The oak variables were one complication too many. Oak might be a good ingredient for crafting certain wines, but it was not a good ingredient in my little terroir-tasting experiment. The juice would not be worth the squeeze. I decided to limit myself, if I could, to wines aged in stainless-steel tanks, not oak. There is no terroir of steel.

So I had defined the limits of my experiment. I would control for the grape variety, its clone, the year the grapes were harvested, and the yeast strain in the fermentation. And I needed these controls across nine different wines made from three different varieties and grown on three different vineyards.

Would this even be possible? And what flavors was I hoping to find that I could pinpoint as *terroir*? I didn't know. But I thought the winemakers at Kenwood Vineyards and Mumm Napa in Northern California could help me.

5

WINE COUNTRY

BEFORE LEAVING TEXAS for Northern California, I wrote to Zeke Neeley, of Kenwood Vineyards in Sonoma County. I knew Zeke actively pursued terroir in his winemaking, so it was my hope that he could guide me through what the industry understood about the phenomenon.

Zeke became the winemaker at Kenwood in 2017, after more than a decade of working for other wineries. Before that he'd done his graduate work at UC–Davis in viticulture and enology. I was drawn to Kenwood because they sourced grapes from many of the appellations (in the United States they're called American Viticultural Areas, or AVAs) within Sonoma County. They curated these into an impressive line of *single-vineyard* expressions. A single-vineyard expression meant that all of the grapes in the wine came from one vineyard—a necessary control for my experiment to taste terroir.

Kenwood Vineyards produces traditional varietal red and white wines. I wrote to Zeke to ask whether I could visit. He invited me out and also referred me to his friend and colleague Tami Lotz, another UC–Davis viticulturist and the winemaker at Mumm Napa in Napa Valley. He suggested she could provide parallel insights through her single-vineyard expressions in sparkling wine.

Zeke's and Tami's pedigrees and their pursuit of single-vineyard expressions made me think I had found the right people for my terroir-tasting experiment. I wrote to them, describing my prospective experiment in detail. They wrote me back to explain why that was a bad idea.

According to Zeke, which Tami echoed:

> No one has really shown how terroir affects the final wine. For the most part, we just know that terroir is one factor that makes wines taste different. But what exactly that difference is, we don't know. I've seen plenty of winemakers claim they know, but they often have little idea of what a control variable might be. And while some academic programs have done extensive studies on terroir, there is always a hole in their experimental design that makes it hard to draw conclusions at the vineyard and winery scale.

This didn't bode well for my experiment. Zeke continued:

> I do believe the location of the vineyard is critically important to the flavor and quality of wine, but I don't think anyone—including Kenwood—can really give you the controlled experiment across three different vineyards and three different varieties that you are looking for. Terroir, and the winemaking process in general, just has too many webs that can impact flavor.

My idea of a laboratory-style, multireplicated, controlled experiment wasn't going to happen. Maybe I should have expected this. I knew as well as anyone that the conditions of a distillery aren't suited for controlled and replicable experiments the way a laboratory is. Why would a winery be any different?

Even though the tasting experiment was unraveling before I'd even set foot in California, I was confident that in whatever form it eventually took, I would still be able to provide some proof that terroir was more than just a marketing buzzword.

So the trip to visit Zeke and Tami was still on. Plus, I had already told my wife, Leah, about my idea for a wine terroir tasting experiment before I had written to Zeke. And what this tasting idea really translated into for her was "I'm getting a long weekend in Sonoma and Napa." So, no matter what Zeke said, we were going to wine country.

* * *

The weather in Northern California is a special blend of warm sunny days and chilly nights, and it was a sunny seventy-five degrees the day we arrived. We spent the chilly night in downtown San Francisco, and then early the next morning we crossed the Golden Gate Bridge and headed north toward Sonoma. Our plan was to spend the afternoon with Zeke at Kenwood Vineyards and the next day at Mumm Napa to meet Tami.

The site of Kenwood Vineyards has winemaking history stretching back to 1906. For more than half of that history it was home to the Pagani Brothers Winery. In 1970, the site was sold to three different brothers, and they dubbed it Kenwood Vineyards. In 1976, the winery acquired exclusive rights to source grapes from the Jack London vineyards, named for the novelist who wrote *The Call of the Wild* and *White Fang*.

The winery is about twenty minutes north of the city of Sonoma, right off Highway 12. The drive from San Francisco follows coastal waters north into immense redwood forests. Then the land opens up all of a sudden to reveal miles and miles of vineyards, hills, and a horizon ringed with distant mountains. Like many Sonoma wineries, the winery sits a good distance from the entrance, nearly invisible. What you do see, though, are rows and rows of grapevines stretching out from the entry road. Given that the winery is tucked away out of sight, a sign announces to drivers not only that does a winery exist here but that tastings are offered daily. A tasting, I thought, is just what I'm looking for.

In the tasting room we were greeted by Alida Westerberg, who was conducting tastings that day. She told us Zeke would be over in a few minutes, so we decided to go ahead and kill some time with a tasting. Alida suggested their "Single Vineyard" flight.

"You read my mind," I told her.

Alida grabbed a handful of wine glasses and put them in front of us before pulling out the first bottle.

"This first wine is a 2017 sauvignon blanc from the Russian River Valley," Alida said, "which is an AVA in Sonoma County. The vineyard is owned and managed by brothers Bill and Ernie Ricioli. They've been growing for Kenwood since 2002."

The Russian River Valley AVA was established in 1983, and its terroir is shaped by its proximity to the Pacific Ocean. The ocean rolls in fog routinely, which keeps the valley's climate cool. The Pacific also created its predominant soil types. Millions of years ago, an ancient inland sea emptied into the Pacific Ocean. The soils it left behind are known as loam: sand, silt, and a lesser amount of clay.

"It's a type of clay loam, quite fertile. It works really well for the clone of sauvignon blanc we use—musqué."

I swirled the glass and brought it close to my nose. *Is that grapefruit?*

"I think I'm getting some grapefruit in here," I said, in a tone that made it obvious I was looking for Alida's affirmation.

"Oh, definitely," she said. "Grapefruit, passion fruit. It's highly aromatic and tropical. That's partly the musqué clone. But the vineyard really allows the unique characteristics of the musqué clone—which is known to have a concentrated tropical flavor—to shine."

"The second wine is a 2015 chardonnay also from the Russian River Valley," Alida said. She rinsed our glasses with water and poured the next wine. "But this vineyard is actually in the Green Valley, which is its own AVA within the Russian River Valley AVA. The Green Valley is the coldest part of the Russian River Valley, and so the fog burns off the vineyards slower. This leads to wines with more acidity and less sugar."

I nosed and tasted the wine. Sour lime and green apple, but still some residual sweetness to balance the acidity and even out the mouthfeel.

"This next pour is also a chardonnay," Alida said, as she went for the third bottle. "But it's a 2016 vintage from the Portola Vineyard, in the Dry Creek Valley AVA. It's to the north of the Russian River Valley and has a warmer climate. The wines typically express more tropical fruit aromas."

Flavor is somewhat subjective. It's common for someone to think they detect a flavor only after someone has told them it's there. But I had to agree once again with Alida. This Dry Creek chardonnay was decidedly fruitier and less acidic than the Green Valley. Part of that could certainly be from the different vintages or some divergence in winemaking techniques of which I was unaware. But Alida attributed these flavor differences largely to the terroir of the vineyard.

"Our last two wines will be reds," Alida said.

I was excited for the shift. I enjoy the crisp flavors of a chilled white wine, but reds are usually bolder. This is largely because white wine is fermented just from grape juice, while red wine is fermented from must—the combination of grape juice and grape skins.

"The first is a 2016 pinot noir from the Olivet Vineyard, which is in the Russian River Valley AVA. The soil is sandy loam, and the climate is characterized by hot days and cool nights. Olivet was one of the first vineyards we sourced from, starting in the 1990s, and it's still one of our best. The 2016 vintage in particular yielded intensely flavorful grapes, due to a dry growing season and moderate yields."

It tasted of plum, blackberry, and cinnamon.

We finished with a 2015 malbec from the Lone Pine Vineyard at the southeastern base of the Sonoma Mountains. This Sonoma Valley AVA is best known for grapes that deliver big-bodied, intensely fruity wines. This 2015 vintage was no exception.

As we finished our last wine, I thought about how foreign these sorts of tastings are in the whiskey world. It wasn't the controlled experiment I originally envisioned, and it didn't tell me anything about flavor-compound variations from one to the next, but it did show how winemakers have always maintained the importance of provenance in their grapes.

* * *

Zeke walked through the doors of the tasting room and greeted us. After I introduced him to Leah and thanked him for meeting with us, he explained what exactly he had in store.

"Like we discussed over email, the type of tasting experiment you envisioned is really difficult to achieve. Wineries just don't operate like laboratories. But I do have two tanks that are in the middle of fermentation that I think you'll be interested in."

We followed Zeke out of the tasting room and up a path to the winery. Inside, dozens of stainless-steel tanks flanked each wall. It was like walking down the corridor of a submarine that smelled really good. Zeke led us down the tanks and stopped at one labeled 122.

"What do the numbers mean?" I asked.

"That's just basically a code so that we can keep track of our tanks and what is in them."

Zeke handed us some wine glasses that he had been carrying and moved to fill up a small pitcher from the sampling valve on the tank.

"OK . . . so what I want to show you are two tanks. They are both 100 percent red zinfandel, and both from the 2019 vintage, so the grapes were harvested just a few weeks ago. Both tanks are from the Dry Creek Valley AVA but from different vineyards."

Zeke opened up the valve and wine rushed into the glass pitcher. I noticed immediately that the color was a little lighter than normal and that the wine was fizzing slightly.

"These wines are still in primary fermentation," said Zeke, "so they taste a little like bubbly, slightly alcoholic grape juice. It's actually delicious. Kids would love it. Too bad there's alcohol in it."

I brought the glass close to my nose and sniffed. Zeke was right. It was full of intense, rich blackberry aromas.

"So do you know which flavor compounds from the grapes make it smell like this?" I asked.

"Not in any definitive way," Zeke said. "I'm sure you understand this, but there are just so many variables with flavor. Obviously, esters are important for fruity notes. But which ones in this wine exactly? I'm not sure. But what I want to show you is just how different this young wine is from the other tank."

Zeke led us around the corner to a second tank labeled 209. He opened a valve and young, bubbly red wine came rushing out into his pitcher.

"Remember," Zeke said, "this wine is nearly identical to the first. It's 100 percent red zinfandel from the 2019 harvest and from the Dry Creek Valley AVA. But the grapes for this second wine came from a different vineyard. The vineyard of this one has heavier soils and cooler temperatures."

We tasted the second wine. It was different—*incredibly* different. Instead of the intense dark fruit, blackberry notes of the first, this was lighter, more floral, and full of bright tropical fruit flavors. It was also paler than the first.

"That is *distinctly* different," I said. "Were these even made from the same clone of red zinfandel?" I asked.

"No, the clones are different. But winemakers usually agree that aside from some very specific exceptions, the influence of the clone is trumped by the influence of the vineyard. So I wouldn't expect the major differences between the two wines to be from the use of different clones. No. These wines were made in identical ways. They were harvested at nearly the same time, and they've been fermenting for the same length of time. The differences in flavor are largely—if not completely—due to the vineyards."

As we sipped, Zeke talked to us about terroir. He said that while it's not something that has been definitively explained scientifically, winemakers universally understand and appreciate how the nuances of a vineyard translate to the wine. They don't know always which flavor compounds come from or are affected by climate, soil, topography, and viticulture. In truth, they often don't which flavor compounds are responsible for the flavor of their wines in the first place. But as they nurture a wine from grape to glass, they think constantly about terroir. Like everything else in winemaking, terroir is a phenomenon that lives at the intersection of art and science.

* * *

The next morning, we woke up early at our hotel in the city of Napa and made our way north to Mumm Napa. Just off the Silverado Trail—which is the less-traveled gateway through Napa Valley than its western counterpart, Highway 29—the Mumm Napa winery is technically within the Rutherford AVA. That said, most of the vineyards from which it sources are farther south, in the Oak Knoll and Los Carneros AVAs. Their own estate vineyard is Devaux Ranch, named for its founder, Guy Devaux, and located in the Los Carneros AVA, south of the city of Napa. In this context "estate" means the vineyard is owned and managed by the winery.

As the name Mumm Napa suggests, they are directly linked to the French-based G. H. Mumm, one of the top five largest champagne producers in the world. G. H. Mumm sent Guy Devaux to the United States in the late 1970s. His goal? Find an ideal winemaking area for the growing of traditional champagne grapes. Like many others who came before him, he settled on Napa Valley. In 1983, Mumm Napa released its first vintage.

Leah and I made our way up the Silverado Trail and pulled into the entry road for Mumm Napa around nine a.m. This winery had a very different look and feel than Kenwood Vineyards. Sprawling patios, awnings, and shaded tables—almost like an upscale restaurant with an undeniable rustic elegance. They hadn't yet opened when we arrived, so it was just me, Leah, and the staff. Mumm Napa is an extremely sophisticated operation, which shouldn't come as a surprise, given they are one of the most successful sparkling wine producers in the United States.

I say "sparkling wine" rather than "champagne" because champagne is sparkling wine produced in the region of Champagne, France, in accordance with the rules of that appellation. You might be familiar with some low-end California sparkling wines that use the word "champagne" (Korbel and André come to mind). This is unfortunate but not a blatant breaking of the rules on their parts. Technically, the United States didn't recognize "champagne" wine as something that could be produced only in Champagne, France, until 2006. Those wineries that were already using the word were grandfathered in and allowed to keep "champagne" on their bottle. Mumm Napa uses the phrase "sparkling wine," a decision I agree with.

Sparkling wine goes through some steps that traditional wine does not. One of those is secondary fermentation in the bottle. The bubbles are created as yeast produces carbon dioxide through fermentation. I wondered, would the nuances of terroir persist through these extra steps? Some of the naysayers of terroir in whiskey say that while grain might possess terroir, it goes through so many manufacturing processes—malting, milling, mashing, fermentation, distillation, and maturation—that whatever flavor there was to attribute to terroir is lost. I don't mean to imply that sparkling wine is more manufactured than traditional wine, but it does go through additional steps to *nurture* the grape to the bottle. So if terroir is maintained in sparkling wine, then that is all the more reason to hope that it exists in whiskey as well.

Tami greeted Leah and me near the entrance, and we followed her into the winery. After a brief tour of their operations, we made our way to the sensory lab, where a table with two different samples was waiting for us.

"I realize you wanted nine or so samples controlled for variety and vineyard, but that is not something we really have available," said Tami.

I assured Tami I'd already quashed any hopes of replicating my experimental setup. But I also told her my experiences yesterday had helped cement my belief that terroir affected flavor, that it was real.

"Oh, it's real, all right," Tami said confidently. "Take a look at these samples here." She gestured to the two sample glasses on the table. "They're both chardonnay grapes from this 2019 vintage. They have both gone through fermentation and will soon be bottled and allowed to undergo secondary fermentation. One comes from a vineyard in the Los Carneros AVA and the other from a vineyard in the Oak Knoll AVA. They are distinctly different, and the only real difference in how they were produced is the vineyard from which the grapes were harvested."

Tami was right. The wine from Los Carneros had notes of green apple and lemon-lime. The wine from Oak Knoll had more ripe fruit. Yet they were the same grape, turned into wine by the same winery, through the same techniques.

"There's a reason winemakers focus so much on the characteristics of a vineyard," said Tami. "It's because it can play such an important part in the flavor of the wine."

* * *

We spent a few more days in California, traveling around San Francisco and the town of Sausalito, just over the Golden Gate Bridge.

A few days in the California wine country was enough to convince me of what winemakers already believed: terroir changed the flavor of wine. I may not have understood all the mechanisms by which it worked, but it was real. I had tasted it. My experiment was not perfect; in fact, it was horribly replicated. But perhaps my dryly scientific approach—multiple replications and quantitative sensory and chemical data—was wrong-headed. When I left California, I took with me an appreciation for how the complexity of terroir is honed and respected by winemakers. They didn't understand it completely, but they had learned how to make it work for them anyway.

Leah and I had a great time exploring, eating, and drinking, but questions kept nagging at me. Winemakers seemed to focus on grapes when it came to the concept of terroir. But did the concept of terroir stretch beyond the grape and the grain? What about the yeast and the oak barrel?

These questions were partly fueled by my growing understanding of terroir, which I now saw as a mechanism by which different species and varieties express flavor through their environment. Terroir was the intertwining of nature and nurture, and any organism will realize interactions between its genetic code and environmental forces. Like grapes and grains, yeasts and trees (and every other living organism) have their genetic code written with DNA, so would the phenomenon of terroir play a role with different yeast strains and oak trees?

One could certainly argue yes. But while we were running around Sausalito and San Francisco, it dawned on me that the concept of terroir in every piece of research or article I had read always seemed to deal with a single type of plant. Grapes, of course, but also coffee beans, tobacco, chocolate, peppers, hops, tomatoes, and blue agave. And what do all of these plants have in common? They are all *crops*.

What makes a crop unique? A crop is not wild. It's domesticated—cultivated and farmed by humans. The environmental conditions to which a crop is exposed to are, at least to some extent, mediated by agriculture. Oak trees are not domesticated, and therefore they are not farmed. The most we can do is manage the forests for sustainability. Likewise, a yeast strain might be propagated for each fermentation, and some yeast strains used for brewing, baking, and winemaking demonstrate the traits of domestication—but they are not farmed. And if oak trees and yeasts are not farmed, they are outside of the domain of agriculture. Both yeast and oak have *provenance*, or a place of origin. But they do not have terroir, at least not using the definition I was developing.

Domesticated grains are the only ingredient of whiskey that fall under the control of agriculture. In fact, they were the birth of agriculture. The domestication of grains enabled humans to remove themselves from the wild ecosystem of hunting and gathering and into the domestic life of towns and cities. My visit to California gave me the information I needed to put my first restriction on my working definition. I settled on this anchoring fact: *terroir is tied to agriculture.* Therefore, from here on out I would focus solely on the grains from which whiskey is made.

6

THE EVOLUTIONARY ROLE OF TERROIR

BOURBON WHISKEY MIGHT be the most famous style of American whiskey, but it wasn't the first. Whiskey, by definition, must be made from grain, and bourbon whiskey uses a majority of corn in the recipe. Corn is the basis of bourbon—it is the primary raw ingredient, and without it (or with a minority of it), you are making something else.

The original American colonists were all transplants from the Old World: Scottish, English, Irish, Welsh, French, Dutch, and German. Diverse as they were, they were all familiar with the same Old World grains. Barley, rye, and wheat were especially common and popular crops in Europe, and the colonists brought these with them to America. While today corn has one of the highest annual production volumes of any crop—and is grown on every continent except Antarctica—it is a New World grain. Barley, rye, and wheat all began their lives no later than 10,000 to 12,000 years ago in the Fertile Crescent of the Middle East. Corn originated at about the same time, but in the valley of the Balsas River in Mexico.

The original distillers were farmers first. In good years, they turned surplus grain into whiskey. While surplus from any grain type would have been distilled into whiskey in colonial America, rye whiskey gained early prominence. This was largely because of rye's agronomic success: Even more so than wheat and barley, rye is extremely hardy. It took to the harsh, long winters and short growing seasons of the American Northeast. Barley and wheat needed generations of plantings and selections by farmers before they had sufficiently acclimated. The immediate success of rye led to surplus rye crops, and, naturally, rye whiskey followed. The first recorded appearance of whiskey in American history comes in 1640, when the director of the Dutch colony of New Amsterdam (which would eventually be renamed New York) called for a distillation of rye whiskey.[1] In the seventeenth and eighteenth centuries, rye whiskey was an unaged spirit. Maturation in oak barrels wouldn't become widespread until the early nineteenth century.

A specific type of rye whiskey became renowned for its quality and flavor—Monongahela rye. This rye was named for the region where it was produced: the Monongahela River Valley, stretching from southwestern Pennsylvania to north-central West Virginia. Through it, from south to north, runs its 130-mile river, which then joins the Allegheny River in Pittsburgh to form the Ohio River. (It's fitting that the Ohio River, so pivotal in the rise of bourbon, starts where the river of Monongahela rye ends.) Monongahela rye was not just celebrated in the new American colonies; even abroad, the style (especially the aged version, known as "old" Monongahela) enjoyed a reputation for its distinct, spicy, and sweet flavors. Monongahela became America's first whiskey terroir.

Near the end of the eighteenth century, the colonists saw the first inklings of the bourbon boom. In 1776, the legislature of Virginia created the Corn Patch and Cabin Rights Act, which offered four hundred acres of land to any settler who built cabins and planted corn in Bourbon County, which at the time encompassed much of modern-day Kentucky and far western Virginia. This act attracted an influx of immigrants and settlers who sailed down the Ohio River from Maryland and Pennsylvania. Whether corn production significantly increased because of this act is debated. But regardless, corn was the draw.

So why corn? If rye was already popular and well known, why wouldn't the 1776 act have been called the *Rye* Patch and Cabin Rights law? The answer is simple—corn grew abundantly and successfully in Bourbon County. While rye could eke out a living, corn was far more prolific. Corn was, after all, originally from Mexico, and it had been brought to the southern regions of the newly founded United States thousands of years earlier. It had spent plenty of time acclimating to the environment. The original whiskey distilled from corn wasn't yet called bourbon, but its distinct sweetness and smooth flavors garnered attention. Bourbon County—now the region called Old Bourbon, which contains within it modern-day Bourbon County, Kentucky—would become America's second whiskey terroir.

A combination of suitable climates, rich soils, and extensive river systems ensured the agronomic success and quality of rye and corn in the Monongahela and Ohio River Valleys. But these terroirs did more than just drive the industry of grain—terroir in effect *chose* the grain because it dictated which type of grain would grow there in the first place. While some corn varieties could grow in the Monongahela River Valley, just as some varieties of rye could grow in the Ohio River Valley, the rye that came with the colonists was already primed for New England climates and especially for the Pennsylvania lands bordering the Monongahela River. And corn was already thriving in the region that would eventually become Kentucky long before being discovered and adopted by European immigrants and settlers.

Why is this? Why did rye come "prepped" for the terroir of New England, while barley and wheat needed time to adapt? Why did corn thrive in the terroir of the southern United States?

To answer these questions, we must go back much further than colonial times. Thousands of years back, to the Fertile Crescent and the origins of agriculture, to the first crop of domesticated grains that would eventually be used to make whiskey.

* * *

The Fertile Crescent extends from modern-day Egypt through Syria, Jordan, and Israel before dipping back down into Iraq and Iran. About 10,000 to 12,000 years ago—at the end of the last Ice Age—our farming ancestors in this region achieved something monumental: domestication of the first crops. Domestication of wild grains led us away from a nomadic hunter-gatherer lifestyle and toward one of settlement, agriculture, and eventually civilization.

So what exactly is domestication? It might be tough to visualize this using grains, but it is a familiar phenomenon. Domestication is the process that turned the gray wolf into the pug.

Domestication is a process of coevolution, where Group A controls—through artificial selection—the growth and reproduction of Group B to the benefit of Group A. In our dog analogy, humans (Group A)—more than 15,000 years ago—selected for desirable traits (obedience, temperament, and loyalty) in certain wolf pups (Group B). Through artificial selection over many generations, the domesticated dog split from the wolf. Humans continued to select for different and desirable traits in a process known as *diversification*. Many, many generations later, domestication and diversification have given us a plethora of dog breeds—both pure and mixed—living happily with us as our loyal companions. We give dogs food, shelter, and belly rubs, and they give us companionship, entertainment, protection, and unwavering loyalty. While grain crops don't give us companionship (unless you just really like hanging out in cornfields), they do give us a reliable source of nutrition. And the domestication of grain crops follows the same story as that of dogs. But instead of selecting for obedience and temperament, farmers and plant breeders selected for traits like larger seeds, hardiness, harvesting success, flavor, and stress tolerance.

The first event of grain domestication was monumental, but wild grains had been a staple part of the hunter-gatherer diet for millennia before domestication. The exact dates are debated, but by at least 23,000 years ago—and possibly as far back as 100,000 years ago—our human ancestors were harvesting, cooking, and eating grains.

The evidence from 23,000 years ago comes from data collected at a site named Ohala II, on the southwestern shore of the Sea of Galilee in Israel.[2] Scientists found starch granules on stone tools used for grinding and discovered burned, ash-covered stones that suggested an oven. In southern Africa, in a Mozambican cave in the Niassa Rift, other researchers again found starch granules on stones that could date as far back as 100,000 years ago.[3] They found no evidence of an oven at this site, and without any evidence of cooking tools, the presence of starch granules might indicate only ancient bedding or tinder for a fire.

In either case, what were these wild grains like, and what brought about their eventual domestication?

* * *

Wild grains, in many ways, deserve their name. Their appearance can appear so foreign to their domesticated descendants that one would never assume they are related whatsoever. While wild teosinte, an ancestor of corn, and modern corn plants share a resemblance, their grain kernels are strikingly different. Whereas the ears of domesticated maize contain five hundred or more naked grain kernels, the ears (rachis) of teosinte contain only five to twelve kernels, each of which is sealed in a stony, nearly impenetrable casing. The kernel and casing of teosinte can survive the digestive tracts of animals, which helps disperse the seeds (a pleasant way of saying that the kernel ends up wherever the animal's excrement does) before the next growing season. Domesticated maize has been selected to discourage dispersion: The kernels, lacking any real protection, have little chance of surviving the gut of a traveling animal. And upon maturity, so many kernels per ear settle into the same location on the ground that the competition for sunlight and resources is too fierce. Ultimately, death is more likely than growth.

Part of this story surely seems counterintuitive. Why would humans create such a feeble domesticated grain? The answer is that humans don't domesticate grains for environmental fitness; in fact, they often do just the opposite. They domesticate grains so that the fitness improves in the context of human goals, which almost always has the effect of reducing environmental survivability and reproduction. Ears saturated with hundreds of unprotected kernels—destined for environmental failure—are much easier to harvest, process, and eat than the desolate ears of teosinte—destined for environmental success—with their small number of rock-hard, encapsulated kernels.

Teosinte is just one example. The wild progenitors of barley, rye, and wheat all share certain common characteristics. They are prone to *shattering* (the agricultural term for seed dispersion), not easily processed because of their tough seed coats, and can go dormant, germinating only when conditions are favorable. The upshot is this: wild grains are a pain for humans to grow, harvest, process and eat. Domestication tamed these annoying wild grains into something accessible. This effort might turn them into feebler plants, ones unable to survive off the farm, but it makes them an ideal part of human society.

* * *

The mechanisms of domestication lie in the concepts of the phenotype and the genotype. The phenotype is the set of observable traits for an organism—what we can see. For instance, the color red is a phenotypic trait for a Pacific Rose apple. Its flavor, too, is a phenotypic trait. The blueprint for the phenotype is the genotype, the sections of DNA that confer these traits. Genes make proteins, and proteins execute the functions that sustain life. Since the Pacific Rose and the Granny Smith have different phenotypes, we could deduce that their genotypes—specifically the genes that control color and flavor—are also different. And we'd be right. The phenotype for color in apples is controlled by genes that respond to light. The genes in the Pacific Rose—because of small variations in the genetic code of DNA—have a greater propensity for responding to light than the genes in the Granny Smith.[4]

Let's say that our neighbor Farmer John plants seeds from an apple that have *segregating* genes for color, meaning that some seeds will contain the genes for the color red and some for the color green. Farmer John, though, can't stand the color of green apples; he prefers the vibrant hue of red apples. Farmer John, being a talented breeder as well as farmer, decides he will plant only seeds from the red apples that grow in the next season. He is selecting for the red phenotype. Over enough seasons of planting only red apple seeds, there is a good chance that eventually the genes for red color will become *fixed*, so that eventually all of the seeds that he plants will turn into brilliant red apples. This process is known as selection—and, in this case, artificial selection.

Our ancestors used the same techniques to domesticate crops. They had no knowledge of DNA and surely didn't use words like phenotype and genotype. What they did know, however, was that when harvesting seeds of the wild progenitors of corn, wheat, barley, and rye, it made most sense to harvest seeds from the "better" plants. What was "better" to the original human domesticators 12,000 years ago? It was often plants that—compared to their neighbors—produced larger seeds that were less prone to shattering. And while most of the harvested seeds would have been consumed for nourishment, some would have been planted to grow the next season's crops. By selecting these larger, nonshattering seeds—whether this selection was conscious (intentional selection by humans) or unconscious (selection by nature because of the human cultivation practices and human-created agricultural environments)—the farmers began to change and fix genes in the population that led to a greater propensity for the desired phenotypes.

These first domesticated crops from 10,000 to 12,000 years ago in the Fertile Crescent are called Neolithic founder crops. The Neolithic period (10,000 to 4,500 BCE) was characterized by an immense transition in human culture. Wandering hunter-gatherers became sedentary agricultural farmers. Successful agriculture depends on predictable crops adapted to the harvesting and nutritional goals of humans. For this transition in human culture to occur, they needed first to domesticate crops.

The noncereal groups of the Neolithic founders included the legumes (lentil, pea, chickpea, and bitter veach) and flax. While I have nothing against peas and have no flak with flax, we can't make whiskey with them. For whiskey, we need cereal grains, and the Neolithic group had three: emmer wheat (*Triticum turgidum*), einkorn wheat (*Triticum monococcum*), and barley (*Hordeum vulgare*). (The species for modern wheat is *Triticum aestivum*.) A few thousand years later, around 8,000 BCE, the farmers of southern Mexico domesticated wild teosinte into corn (*Zea mays*), called maize in much of the world. The domestication date of rye (*Secale cereale*) is a little trickier to pinpoint, as it often persisted as a weed in fields of domesticated barley and wheat, but by around 6,000 BCE—and maybe a few hundred or thousand years earlier—rye had also been domesticated. These regions of domestication like southern Mexico and the Fertile Crescent are called *centers of origin.*

A center of origin is the place where a crop was first domesticated, but it is more than just a physical slice of land. The center of origin is also the first terroir that sorted genetic variations in the wild varieties through natural selection—survival of the fittest. Then, through artificial selection, humans domesticated those wild varieties, selecting for the characteristics that they desired and that could thrive in the local climate, soil, and topography. When the grains were transported by humans away from their center of origin, they adapted to new terroirs and continued to evolve.

The term "centers of origin" was coined by the prominent Russian geneticist and botanist Nikolai Vavilov in the early part of the twentieth century. After collecting and analyzing seeds from around the world, he realized that the site of domestication was also home to the maximal genetic diversity for a crop. This is because wild grains are more genetically diverse than their domesticated varieties, and the farther a domesticated crop is carried by humans from its center of origin and subjected to deliberate breeding, the more genetic diversity is lost and the more likely its genes will become fixed.

All this said, the barley, rye, wheat, and corn that we know and grow today are very different than the original domesticated crops. While domestication is the original transition of a wild variety to a crop variety, diversification is the postdomestication selection process that facilitates the continued change and adaptation of the crop. Domestication laid the foundation, but diversification paved the road that led—and still is leading—to a huge diversity of grain varieties. Diversity among varieties of a grain species is apparent and recognizable by all people: nobody would mistake popcorn for sweet corn, yet they are varieties within the same domesticated species—*Zea mays*. What is responsible for this diversification? And how can this diversity change the flavor of whiskey?

* * *

Remember the wolf and the pug? We know the pug is a domesticated variety of the wild gray wolf, but is it plausible that that first instance of domestication led straight to a pug?

In reality, the first domesticated wolves were probably just wolf lineages that were less aggressive and more comfortable with humans. From there, humans began to select puppy wolves for more acute hearing (to alert that predators were near) or maybe a heightened sense of smell (to assist during a hunt). The domesticated wolves were selected and bred for other environments or purposes: a tolerance to the cold, a propensity to retrieve, or an innate loyalty and desire to protect their human masters. And from there the domesticated wolf diversified into all the dog breeds we know today. The pug is a result of domestication followed by extensive diversification. The same thing happened to grains.

Diversification has three stages that occur simultaneously. Any of these can lead to a new crop. The basis for diversification—just as in domestication—lies in the same concepts of genotype and phenotype. Stage 1 occurs at the center of origin, and diversification in this stage is largely driven by human selection for yield. This stage overlaps with the domestication event itself and concludes with the emergence of fully domesticated crop varieties.

Stage 2 occurs when the newly domesticated crops spread from the center of origin. Humans took seeds with them when they migrated to new lands, exposing them to new environments, new terroirs. These seeds were not genetically identical and therefore grew into plants with varying phenotypes. The environments of the new terroirs influenced the phenotypes through selection: some plants would be more successful than others. Humans would artificially select for those newly diversified varieties that could thrive in the new environments and exhibited desirable qualities, like those that were easier to process and cook. Flavor, too, became important in Stage 2. Domestication was largely—if not solely—about a sustainable source of food. Diversification brought new and desirable flavors to be discovered and selected for. Many of the heirloom varieties of corn, wheat, barley, and rye that are still around today arose in Stage 2.

Stage 3 happens through the deliberate breeding of existing varieties into new and improved ones. Improvement is subjective and is connected to the desires of the breeder, but in general with food crops the goal is to increase yield, uniformity, farming success, and quality. Deliberate breeding is often associated with hybrids, in which one domesticated variety (Parent 1) is genetically combined with another domesticated variety (Parent 2). Usually, Parent 1 contains desired traits that Parent 2 does not have, and vice versa. Most techniques within Stage 3 are products of modern times, with some exceptions (the hybrid breeding of figs, for instance, has been practiced for more than 11,000 years).[5] Some of the more significant—as well as notorious—breeding advances within Stage 3 are those of genetic engineering. This can include the transfer of foreign DNA into a plant, which is how genetically modified organisms (GMOs) are created. The idea here is that the foreign DNA (often from a bacteria or fungi) can confer some type of pesticide and stress resistance. Another form of genetic engineering is known as gene editing, in which the plant's own native DNA is edited (not mixed with foreign DNA) in order to control the expression of some gene. The CRISPR/Cas9 system is the most famous recent example of a gene-editing technique.

The important takeaway is that domestication and diversification have led to thousands of different varieties of corn, wheat, barley, and rye. But every variety can technically trace its lineage back to a single wild progenitor. Human efforts in artificial selection produced the first domesticated varieties. From there, humans transported them around the globe, growing them in any environment that was suitable. The new environments and new selections bore new varieties. This process has been going on for thousands of years, and because of it we have thousands—even hundreds of thousands—of varieties of the four major whiskey grains: corn, wheat, barley, and rye. Each variety has its own unique phenotypes: kernel color, size, stalk height, climate preferences, and so on. Each variety is distinct. And just as with grapes and wine, each grain variety has its own flavor.

The genetic and phenotypic diversity of the grain varieties is indeed immense. But of the thousands of varieties on the planet today, many exist only in the fields or barns of a single farmer. Many other varieties exist only as a handful of kernels tucked away in a box or bag in a single seed bank. This is because modern agriculture has sacrificed genetic diversity in favor of yield and uniformity. We buy seeds from only a few major seed companies—such as Bayer (formerly Monsanto), Syngenta, and DuPont—which typically breed new varieties using previously utilized germplasm (to which they own the rights), to protect their intellectual property. This means that the varieties commercially available today—at least those in widest production, which are those from the major seed companies—are often relatively similar, at least compared to the total diversity of all varieties within a given species.

What happened? Why did modern agriculture strip away thousands of grain varieties from working production? Why did it take away the equivalence of wine grape diversity in grains and deliver in its place the monotony of the Concord table grape?

7

THE RISE OF COMMODITIES

"THE KEY TO good food is to decommoditize it."[1] This quotation is from Bob Klein, the founder of Community Grains in California. Community Grains is devoted to supplying the general retail market with pasta made from heirloom wheat varieties grown on specific farms in California. Heirloom grain is loosely defined, but usually it means an older variety that was bred, selected, and passed down generation to generation outside of commercial breeding efforts. Unlike most modern commodity varieties that were bred solely for yield, heirlooms were often bred for flavor. The commodity wheat that large pasta makers buy is a mix of modern varieties bred for yield, and so their flavor is usually dull compared to heirloom grains. But by segregating varieties and farms, the pasta from Community Grains carries the distinctive flavors produced by terroir. This separation of variety and farm does not happen with commodity wheat.

Community Grains actually goes further than preserving variety and farm. They offer "23 Points of Identity" for the pastas they sell. Essentially, these identities allow the customer to trace the pasta from "seed to table," through twenty-three important details about its farming and processing. Beyond variety and farm, some items that are relayed are harvest date, soil management practices, irrigation use, and how the flour was stored.

If the key to good food is to *decommoditize* it—and such an idea can be extrapolated to whiskey—then what exactly is the problem with commodity grain? Is commodity grain destined to make bad, boring, bland whiskey?

The answer, perhaps surprisingly, is no. Generic commodity grain can produce delicious whiskey. And millions and millions of gallons of whiskey are made this way annually and sold to happy consumers. The diversity and quality of whiskey brands and flavors are as high as they have been since before Prohibition.

But is the whiskey industry missing out? If distillers used grain that was effectively decommoditized, grain that reflects the terroir of the place in which it is grown, could we capture something else, something beneficial, something important?

* * *

Agricultural commodities—like coffee, cocoa, fruit, sugar, and grain—are called "soft" commodities. Soft commodities are grown; hard commodities—gold, oil, and extractive natural resources—are mined. But all commodities are fungible, meaning they are interchangeable and equivalent. A ton of food-grade yellow dent corn kernels from one farm is the same as another from a different farm. Treating grains as commodities and trading them in markets is practical. In theory, the commodity system financially protects the farmer, the buyer, and the consumer.

A standard commodity contract includes the commodity grain type (corn, wheat, barley, rye), the quantity, and a price based on the quality of the grain. There are also different classes for each grain type, like hard red wheat versus soft red wheat, and there are quality grades for grain lots, usually specifications set by governments. These starting specifications are typically test weight (a measure of how much volume a given weight of kernels occupies), the proportion of damaged and broken kernels, and the percentage of foreign materials, like rocks and sticks.

In the United States, most grains are graded on a scale set by the U.S. Department of Agriculture. Barley and rye are graded on a scale of one to four; corn and wheat, one to five. The lower the number, the higher the quality and the more money it's worth. Once the grade is set for a load of grain—whether it's being delivered in a large bag, semi truck, railcar, or even shipping liner—then that load is completely interchangeable with all other loads with the same grade. Regardless of terroir—regardless of the variety or farm—a load of Grade 2 yellow dent corn is interchangeable with all other Grade 2 yellow dent corn loads.

Unlike corn or wheat or rye, barley is often malted before it's used in whiskey making. Compared to raw grain, malted grain has enzymes that convert starch to sugar and a range of new flavors. Contracts for malting barley usually have higher quality specifications than what the USDA dictates for commodity barley; for instance, they will often specify certain protein levels and germination rates. Low protein is necessary to ensure sufficient starch (which is indeed desired, as the starch will eventually become alcohol) and efficient lautering (the process whereby mash is filtered and becomes wort). But just as with two loads of commodity barley with the same USDA grade, a load of two-row malting barley that meets the malthouse specifications is interchangeable—regardless of terroir—with all other two-row malting barley loads of the same specifications.

When I asked a large, international maltster that we at one time worked with which varieties of barley they used, he simply directed me to the thirty or so varieties approved by the American Malting Barley Association, Inc. "Any of those, I guess," he said.

When I asked where the barley farms were located, he said, "Oh, you know, around Minnesota . . . and also North Dakota, Montana, and Canada."

There are exceptions: MillerCoors apparently has proprietary barley varieties that they mandate for their contracts. But by and large, most malting barley is still handled, traded, and sold as a commodity.

The quality specifications set by governments and malthouses are important parameters. Whiskey distillers care about test weight, foreign material, broken kernels, and protein levels. But there is an important quality specification that is not measured, recorded, or rewarded when commodity grain is traded: flavor. Compared to wine grapes, this is an important differentiator. When wine grapes are traded—usually through contracts between a grower and a winemaker—what they are after is flavor. And through the combination of grape variety, location, and viticultural techniques, terroir drives this. This is why such specifications are also laid out in wine grape contracts. Grape growers are rewarded for flavor in ways that grain farmers are not. In a contract between a vineyard and a winery, the price of the grapes is a negotiation. Those vineyards with a reputation for flavor can demand premiums far above the average.

But when commodity grain is traded, the situation is drastically different. Distillers (or any other end users for that matter) do not try to impress farmers with courtship displays, financial or otherwise. The severing of the distillery from the farm that largely took place after Prohibition has meant in most cases that whiskey distillers do not know from exactly which farms their grain comes. And personal relationships between farmers and distillers are increasingly rare. Unlike vineyards, grain farmers have little control over the price of their grain. Baseline prices are instead set by commodity markets, traders, and grain dealers. Different grades demand different price points, but flavor is never a quality parameter in any of them. So the question is, then, why isn't flavor in grain—flavor that can come from terroir—valued?

* * *

Up until the 1880s—when the first large-scale seed companies were established—nearly all crop varieties were heirlooms.[2] Flavor would have indeed been an important selection trait for these heirloom varieties, as they would have been used primarily as food. And it wasn't uncommon for different varieties to be associated with certain regions, usually because they were bred and selected by the farmers themselves. In the 1800s, thousands of small-scale distilleries dotted the towns and cities of the United States, and they would have used these flavorful heirlooms. Without an effective grain transportation system and given their relatively modest size, these distilleries sourced grain from their local farmers or sometimes were farmers themselves. And being small scale, the local farmers could have stored sufficient excess grain on their own land or in local communal storage facilities. The whiskeys made from this local heirloom grain would have carried a signature of terroir.

This changed in the twentieth century, however, for multiple and overlapping reasons involving grain elevators, steak, and white bread.

Grain elevators are facilities built to store grain. They were introduced in the middle part of the nineteenth century. The word *elevator* refers to how the grain is elevated—through some type of augering system—from the ground level as it's offloaded from truck, train, or ship and up into the silo. They come in many shapes and sizes, but regardless of their organization, they usually cater to the commodity grain market, and typically they are not set up to effectively segregate grain. This is not always the case: some smaller-scale grain elevators are perfectly adept at keeping certain grains separate from others. But most grain elevators purchase large volumes of grain from many farmers, blend everything together, store the grain for extended periods, and then sell to the various industries—feed, food, and fuel—that use grain. And after the development of the grain elevator, a surplus of grain quickly followed.

The early twentieth century saw the introduction of synthetic chemical fertilizers and hybrid grain varieties, which boosted crop yields hugely. For the first time in history, there was a surplus of grain so great that it could increasingly be used as animal feed. Animals have been farmed since their domestication some 12,000 years ago. For much of the history of animal husbandry, livestock ate by foraging wild or domesticated grasses. In the United States, this was the case for virtually all livestock through the first half of the nineteenth century. Grain was fed to animals in some cases, but it would have largely been the leftovers after human consumption. Wheat, barley, rye, and corn were simply too valuable as human food to waste on animals.

However, with a grain surplus, many mills switched over to processing grain—mainly corn and barley—into animal feed. This in turn meant Americans could farm more animals than ever before, and our diet shifted accordingly. Animal protein (milk and cheese included) wasn't just expected at every meal; it moved to the center of the plate. And because chickens, cows, and pigs didn't have any discernable preference for flavor in their feed grains, the market had less concern for flavor. Growers and breeders began selecting purely for yield.

But not everyone could afford steak. Bread was still the staple food in many people's diets even as we began to supply the market with more meat. For much of human history, bread was produced from whole flour. Only the rich would have been able to afford white bread, which was seen to be more pure and superior to whole-wheat bread. But as roller mills were introduced in the late nineteenth and early twentieth centuries, white bread became available to all. The new mills made it possible to separate a kernel of grain into its three constituent parts: the bran, the germ, and the endosperm. The bran is the protective seed coat, and the germ is the embryo, the baby plant. They are packed with much of the flavor compounds that a grain kernel can deliver. But both also contain lots of fatty acids, which cause whole-grain flour to spoil relatively quickly. The endosperm, on the other hand, is the food reserve for the baby plant, and it is packed with carbohydrates in the form of crystalline starch molecules, which are not prone to spoilage. The endosperm is also relatively flavorless. If the bran and germ can be removed during milling so that the endosperm is all that is left, then the flour market could easily create white flour that would be shelf stable for years rather than just months. Shelf stability thus meant that local sources of wheat were no longer necessary. Wheat could be bought from all over, milled into white flour, packaged, and transported. White flour was in demand, and the commodity grain trade supplied the wheat. It might have been low in flavor, but nothing that butter or jam couldn't liven up. And since this endosperm-rich white flour was more or less flavorless, the flour market did not reward farmers for growing more flavorful wheat. Today, we call this flour made only of the carbohydrate-rich endosperm "refined flour" or "all-purpose flour."

So, ultimately, we stopped focusing on flavor in grain at the same time the markets found new customers who didn't care about flavor. You might still ask (as I did): Why didn't whiskey distillers insist that the farmers keep growing the traditional, flavorful varieties?

It's hard to pinpoint a specific reason. For one, as grain yields increased and a surplus was created, the price of grain dropped. Many distillers would have welcomed the lower prices, even if it changed the flavor. And it's likely that these new high-yielding grains didn't have their flavor diluted overnight. Only in the last decade has the scientific literature revealed that when seed genetics are tuned for yield, it can inadvertently suppress flavor.[3] But it isn't an overnight reduction. It's slow and gradual. Further, the whiskey distilling industry experienced rapid consolidation and growth in the twentieth century after Prohibition, and it became increasingly difficult for distillers to source all the grain they needed from local farmers. While most farmers could store some grain on their own farm or at town granaries, the sheer volume of grain demanded by the newly consolidated distilling behemoths was really only achievable with the large-scale, commodity grain elevators. Any flavor nuances from terroir would have been lost in these grain elevators, blended into uniformity. Whether the grain comes through a commodity contract, a food contract, or a malt contract, once it reaches the elevator, it's shuffled into a silo and mixed with all the rest, all the other varieties and terroirs that match the superficial specifications of the USDA: species, grain class, and grade.

* * *

Throughout the twentieth century and most of the twenty-first, grain that was of acceptable USDA grade has been good enough for distillers. But there's a movement spreading. Whiskey distillers more and more are starting to question the status quo of grain selection. If terroir is so important to winemakers, if they place such an emphasis on variety and vineyard and the flavor of their grapes, then why don't we?

Many large and small distillers, craft outfits and the big boys, in the New World and the Old World, are starting to wonder the same question: Can we tap into terroir? Can we capture, highlight, and promote the nuanced and distinct flavors of soil, climate, topography, and specific varieties? The twenty-first century has seen a rapid proliferation in the number of distilleries worldwide. This was spurred by many factors—the craft-beer industry, the foodie movement, and the locavore mentality. As the number of distilleries grew, the number of distillers that joined the industry—from many different walks of life—has consequently increased.

Early in our careers, many of us saw the stark difference between how winemakers select grapes and how distillers select grains. And the marketability of "drink local"—originally meaning brewed or distilled locally—is that much more powerful when the grains are also, well, local. We can make good whiskey with commodity grain. But can we increase the diversity of flavor by using grains from specific farms and of specific varieties?

Farmers do not make very much money selling commodity grain. It's not uncommon for farmers to invest more to plant, grow, and harvest a bushel of corn than they actually profit once they finally sell it. So why do farmers grow commodity grain at all? One reason, and perhaps the most important, is that commodity grain is the proven model. While growing for the commodity market is far from perfect, it does provide a guaranteed buyer of the grain once harvested, making agricultural loans easier to secure. Another reason is government subsidies. Subsidies come in two packages: direct payments and crop insurance. The former offsets profit losses when commodity prices are down, and the latter will pay back the farmer if the crop is lost or if the yield is abnormally low. Further, even if farmers did grow grains with the goal of tapping into unique flavors, how do we effectively segregate and store grain of specific varieties and farms so that we don't blend away terroir in commodity grain elevators?

The wine industry provides no insight here. Once harvested, grapes are delivered to the winery to be crushed and processed. A winery presses grapes to create wine only once a year. Whiskey distilleries mill grain to create beer and whiskey routinely throughout the year. Because grapes are not stored, it is immensely easier (often necessary logistically) to segregate varieties and vineyards and maintain terroir. If grapes could be stored and transported like grain, wine too would have likely gone the route of grains, and we might not have had much more than a generic red and a generic white grape. Would this mean these generic grapes would create bad wines? Not necessarily. But it would mean that—just like what's happened with whiskey—flavor nuances from different grape varieties and terroirs would be lost to the blending nature of these imaginary grape elevators.

Luckily, we don't need to look to the wine industry for an answer. The answer actually already exists, even if most distilleries don't use it. It lies in a system devised largely to segregate different varieties for seed production and distribution. It's called identity preservation, and it begins with seed certification.

The seed certification system is designed to maintain and preserve the genetic identity of a crop variety. Remember the "23 Points of Identity" from Community Grains? When farmers buy certified seed, they are guaranteed that the seeds are all genetically identical. This ensures the consistency of traits like yield, hardiness, pest resistance, or flavor.

There are four classes and progressions within seed certification: breeder, foundation, registered, and certified. *Breeder seed* is—appropriately—controlled by the breeder. If a variety of breeder seed appears to have commercial promise, then those seeds are planted, and the progeny harvested is known as *foundation seed.* At this point, certain state and federal laws dictate standards for genetic purity and identity. If the variety is promising enough commercially, the foundation seed is reproduced to create *registered seed.* This seed is then planted, and its progeny are known as *certified seed.* This certified seed is what is sold commercially to farmers.

Identity preservation starts with a certified seed variety. It then calls for all equipment—planters, combines, augers, elevators, and storage bins—to be thoroughly clean and free of any contaminating material or genetically different seed or grain. (The difference between seed and grain is simply in its purpose, as they are both biologically the same—they are both kernels. But seed kernels are planted, and grain kernels are harvested and sold for end-use products.) Identity preservation is most concerned with genetic purity, which means segregating different grain varieties.

Identity preservation captures terroir. Or at least it could, in theory. And for small distilleries that work directly with farmers, it's relatively easy to use identity-preserved grain. All it requires is a grain-handling system to fill the two-thousand-pound bulk bags that are the industry standard. If the distillery wants to use identity-preserved malted grain, the situation becomes a bit more complex, as they'll need to find a maltster within a reasonable distance. But that too is possible with the recent rise of craft malt houses.

Large distilleries don't traditionally use identity preservation. This is chiefly because of their volume and storage needs. For a large distillery to identity preserve enough grain to supply their grain volume between harvests, they'd need to work with farmers covering thousands to tens of thousands of acres and secure enough grain silos to identity preserve hundreds of thousands to millions of bushels. This isn't impossible, but it is more expensive and more difficult logistically than simply buying the blend of varieties and farms that are constantly available from commodity grain elevators.

Before making efforts to capture and highlight terroir, large distilleries would need to be convinced it's something real and worthwhile and something that commodity grain elevators can't deliver. To be honest, even most small distilleries would take some convincing. Although identity preservation is much easier in two-thousand-pound bulk bags, there are still the financial and logistical issues of finding the right farms, harvesting the identity-preserved grain separately, and storing the bulk bags between harvests.

In addition to these challenges, most grain farmers have spent their lives selling solely to the grain elevators that serve the commodity market. For farmers to grow, harvest, store, and sell grain that maintains the identity of terroir and is destined for whiskey as opposed to a commodity grain elevator, distillers would have to reward them for their efforts. Distillers would have to reward them for flavor, and the offer would have to be more attractive than what commodity grain offers.

So, the question really becomes—for both small and large distilleries—is it worth the effort? Do the flavor nuances among grain varieties and farms translate to whiskey, or does the distilling process strip away any distinctions? Maybe it's the case that each grain species—corn, barley, rye, and wheat—will always deliver the same flavor profile regardless of terroir.

But let's assume that terroir does affect flavor in whiskey. Which aspects are most important? Is it the grain variety, the farm, the climate, the agronomics, or that elusive human element? When all is said and done, how much does the flavor of a place—that palpable perception of provenance—really change whiskey?

To answer these questions, I needed a guide. I needed something that would anchor the underlying science of all these questions and provide concrete insight into how specifically terroir affects the flavor of whiskey.

What I needed was a roadmap.

PART II

A ROADMAP TO TERROIR

8

A TEXAS TIC-TAC-TOE

SAWYER FARMS, THE farm from which we source all our grain, is about one hour south of our distillery in Hill County, right on the border of North and Central Texas. The soil there is mostly Houston black clay, which is found mainly in the Texas blackland prairies, extending from the Red River at the border of North Texas and Oklahoma all the way south to San Antonio. It's a rich soil capable of supporting many crops. Before we met, John Sawyer used it to farm thousands of acres of corn and wheat for the commodity market. After our partnership formed, he added rye and barley.

I distilled my first legal batch of whiskey at TX Whiskey in February 2012. This first distillation happened at our first distillery at 901 W. Vickery Boulevard in Fort Worth. In the months leading up to that first distillation, we searched everywhere for a reliable and local source of grain. I worked mostly with the distillery's proprietors, Leonard Firestone and Troy Robertson. We didn't discuss the idea of terroir—much less appreciate the complexities of it—but we did understand the idea of provenance. We believed whiskey required a sense of place.

About twenty minutes north of our distillery was Attebury Grain, a large regional grain elevator that served the commodity market. At any given time, Attebury held millions upon millions of bushels of yellow dent corn and soft red winter wheat, two of the grains from which we would make our wheated bourbon. Remember, bourbon by law must contain at least 51 percent corn in the grain bill, and most bourbons make up the other 49 percent with more corn, rye, and malted barley. But some bourbons use wheat instead of rye, and those are called wheated bourbons. Some of the most well-known "wheaters"—as the whiskey geeks call them—are Maker's Mark, W. L. Weller, and Pappy Van Winkle. When it came time for our first distillation, we had local wheat in spades but no local rye, so we opted to make a wheated bourbon. Five years later, that first 2012 batch became the first bottling of TX Texas Straight Bourbon.

We were pleased with our local source of corn and wheat. But soon I began to yearn for more control and detail on the "local" aspect.

Attebury was a large, regional commodity market grain elevator, and that meant the only details they could really guarantee was that the grain we bought was grown in Texas. That narrowed it down to about 130 million acres. They couldn't tell us which region of Texas, much less the specific farm or even the grain varieties.

This didn't sit right with me. It wasn't that the elevator's grain was low quality, but I couldn't help but feel that we could better control the consistency and the quality of flavor if we could always source the same varieties from the same farms. I didn't think that our "local" grain ought to be grown hundreds of miles away in the Texas Panhandle across dozens of different farms. There had to be a better way to source local—truly local—grains.

I expressed my concerns to Attebury, and they understood completely. In fact, they wanted to help. We were, after all, a tiny client for them, filling only a handful of small two-thousand-pound boxes per shipment. The time and energy these took to load and deliver was almost more trouble than it was worth. George Gurganus, Attebury's grain originator, offered to introduce me to some of their farmers who might be interested in selling us grain directly. One day in 2014, George brought a group of farmers to tour our distillery.

This wasn't my first time giving a distillery tour to farmers. I had done it quite regularly since 2012, largely as a favor to Attebury Grain. As I said, we were more trouble than we were worth, except in one key way. Hundreds of farmers sold grain to Attebury Grain, but Attebury wasn't the only grain elevator in town. The company tried all sorts of ways to lure farmers to them: taking them to lunch, throwing them farmer appreciation parties—and sometimes taking them to a whiskey distillery that used their grain. we had made a more or less unspoken agreement with Attebury: we would pay them the same price for grain as other buyers, even though it was more work to load and transport the grain, and as a thank you, we would open our distillery doors to them whenever they asked.

By 2014, I had given dozens of tours to George and Attebury's farmers. The group on this particular tour was composed of ten farmers, along with George. They all farmed within a hundred-mile radius of Attebury, and George had offered them a quick day trip to tour our distillery. Nine of the farmers had the same reaction as every other tour-goer, equally intrigued and surprised by the complexity of whiskey making. But one farmer was much more observant than the others. He seemed already to understand the process. When he tried the new-make bourbon—the unaged, straight-from-the-still, 135-proof white dog spirit that tastes very different than mature bourbon—he didn't wince or gasp like the others. He took his time to experience the flavors, and the strength didn't seem to bother him. I knew he knew whiskey.

Near the end of the tour, I found him opening up our bins to inspect our grain. As I walked up, he turned to me and confidently said, "I can do better." This was John Sawyer.

As we started chatting, it was apparent that George had already talked to him about my concerns and goals. "So, George tells me you want to source grain directly from a farmer," he said to me.

"Ideally, yes," I said. "Working with George and Attebury has been great for us. But their operations are set up so I can't control the grain varieties or which farms we source from. All I can specify is that the grain is grown somewhere in Texas."

It turned out that John had worked with winemakers in the 1990s when he was an agricultural loan specialist. He was familiar with the importance they placed on terroir, varietal, and vineyard. Before he visited TX

Whiskey, he had already been working on plans for a distillery of his own. He had matriculated at Moonshine University, in Louisville, Kentucky, which trains would-be distillers in the arts of whiskey making.

John told me he'd been thinking about making whiskey himself using grains grown on his farm, a sort of single-estate whiskey, if you will. But as much as he liked the idea, he was a farmer, not a distiller.

"Farming is what I do," he told me, "it's what I grew up with, and I'm good at it. So . . . if you're still looking for a farmer to work directly with, then let's say you and I talk more? I can provide you with specific varieties that will only come from my farm, and I'll segregate them for you at my grain elevator."

"*You have your own grain elevator?*"

He laughed. "I have my own grain elevator. The silos are much smaller than the ones at Attebury, of course, but they'll hold much more than you would use from harvest to harvest. And I have enough to where I could—without muddling my current operations—designate a few silos to hold your identity-preserved grain."

I had talked to many farmers by now, but the conversation always ended at the same place: "Where are you going to store the grain if you buy directly from me?"

We had some storage capacity at our distillery, but only enough to keep about a week's worth of production grain. But John could hold more grain in one silo than we'd ever need to use from harvest to harvest. His elevator, which he named Apex Grain, was a regional granary that bought grains from farmers near his operation in Hillsboro. John then sold the grain to Attebury in Saginaw, among others. Apex typically received, mixed, and stored grain in the usual way, blending away any nuances from unique varieties or farms. But he would store ours separately—the opportunity for identity preservation that would turn the grain elevator into a bastion of terroir.

"We have test plots of different corn varieties growing right now," John said, "and we recently harvested wheat-variety test plots. Why don't I send you some of those wheat varieties now and then some of those corn varieties after harvest. You can test them, let me know which you like the best, and then we can discuss growing and storing those varieties specifically for you."

This was the beginning of the most important working relationship in TX Whiskey's history. It was a relationship that would effectively sever both of us—our distillery and Sawyer Farms—from the commodity grain market. Over time, it would turn our various expressions of whiskeys into "single-farm" whiskeys—a term analogous to "single-vineyard" wine—meaning all of the grain that made its way into the bottle came from one single farm.

In 2015, we distilled the first batch of whiskey using Sawyer's yellow dent corn and soft red winter wheat. From the first sip of the white dog, I knew we'd captured something specific. Compared to the whiskey we made using commodity grains, the whiskey from Sawyer Farms grain had more fruity and floral notes and cleaner flavors of sweet corn. Just as I had tasted terroir in the wines at Kenwood Vineyards and Mumm Napa, I could taste the flavors from Sawyer Farms grain.

What was responsible for this change in flavor? The environment of Sawyer Farms was partially responsible, of course. But it was more than just that. Identity preservation meant that we would work with only *one* variety of yellow dent corn and *one* variety of soft red winter wheat. For variety selection, we considered agronomic yield, like any commodity grain farmer would, but we also considered flavor.

To assess the flavor of raw grain, distillers put a small amount—fifty grams or so—into a glass snifter, cover it with a watch glass, and pop it into the microwave for fifteen seconds. Then they remove the watch glass, and the aromas of the grain waft up from the glass. In the raw grain, we were looking for aromas of toasted nuts, sweet corn, coconut, and/or sweet spice. If we detected any of those notes, then we hypothesized that the variety was higher in the precursor flavor compounds present in oils, proteins, and lignin. By microwaving the grain, the precursors had some chance to be degraded into their flavor compounds. The most promising varieties from this aroma test were processed in the lab into mash, beer, and new-make whiskey, so that we could make our ultimate decision based on the flavor of the whiskey itself.

* * *

The varieties we chose were undoubtedly "modern varieties," none of the romantically named heirlooms like Bloody Butcher corn or White Sonora wheat. New varieties of corn and wheat are named for the seed company that made them, like LA841 Terral wheat and D54VC52 Dyna-Gro corn.

Since we had chosen modern varieties, this meant we had to reckon with the fact that they had been bred and selected for yield, not flavor. This is the case for nearly all modern grain varieties. You might assume, then, that modern varieties bred for yield might not taste as good as the older, heirloom varieties. In many cases, this is true. Consider heirloom tomatoes. Their flavor is more intense and diverse than the large, perfectly round, ruby-red modern tomatoes.

Perhaps "tastes better" is more subjective than objective, but the scientific research does bear out the flavor difference. Modern high-yield crops have less intense and diverse flavors than heirlooms. Harry Klee's lab at the University of Florida proved that as breeding efforts focused on yield—specifically the number of fruits per plant and the size of the fruit itself—the concentration of flavor compounds inadvertently declined. The hypothesis is that as we breed for yield, some of the genes that produce flavor compounds are rendered inactive through DNA mutations, deletions, and insertions.[1]

Modern breeding programs have avoided or ignored most of the available varieties in a species to focus instead on what they call "elite lines," varieties with a track record of agronomic success. On the surface, this seems like a sensible approach. Why not use the best varieties as parents for the next generation of offspring?

The trouble is, it has also led us to consider only a relatively small number of varieties as parents for new generations, and then varieties from that next generation are themselves used as parents for the next generation. This way of breeding has reduced the genetic diversity of our crop species. And less genetic diversity coupled with the inadvertent effects of breeding for yield means less flavor diversity.

It's important to realize, though, that while the vast majority of whiskey is produced with modern grain varieties that were bred for yield, breeding for yield doesn't necessarily create grain that "tastes bad." The science generally shows that the concentration and diversity of flavor compounds only *decreases*. So it tastes not *bad* but *less*. This reality led me to an important question.

Given that modern varieties potentially harbor so much less genetic and flavor diversity than heirlooms, is it possible that they do all just taste the same? Regardless of breeder, variety, or farm, do all modern grain varieties—within their respective species—deliver the same flavors to whiskey? If the answer was yes, then there would be no reason for large-scale whiskey distilleries (including us) to make the logistical and financial effort to leave the commodity grain system. While heirloom grains do supply a niche in the whiskey industry, on the farm they can yield only half as much as modern varieties. The reality is that the vast majority of farmers and distilleries—especially those producing tens to hundreds of thousands of barrels annually—can't drastically sacrifice yield (especially a 50 percent one) for flavor. We need both yield and flavor.

Grain flavors in whiskey come through multiple pathways. Concentrations of sugars, amino acids, and nutrient compositions will change the production of flavor compounds by yeast during fermentation. In processing, precursor compounds in grain can undergo reactions like thermal degradations, oxidation, the Maillard reaction, and Strecker degradation, which deliver flavor compounds that survive in the finished whiskey. And secondary metabolites ("secondary" means these compounds are not required for the organism to grow and reproduce but do increase the chance of survival, such as through chemical defense) produced by grain can theoretically influence whiskey flavor directly versus as a precursor for metabolic or chemical reactions.

My thought—or maybe my worry—was that if there is limited genetic diversity in the relevant biosynthetic pathways among modern grain varieties, then terroir might not matter. Personally, I had enough proof after the first distillations using Sawyer Farms grain that there was something *there*. But maybe it was some phenomenon besides grain variety or farm—maybe it was the storage and transportation conditions, for instance. But if the flavor difference was indeed terroir's influence on the production of flavor compounds, I still had to test this. I wanted to know definitively whether terroir changed grain-derived flavors in whiskey when modern varieties were used. I decided to set up a laboratory experiment to see whether I could prove scientifically the effects of terroir on whiskey.

We use yellow dent corn and soft red winter wheat in our TX Texas Straight Bourbon, with yellow dent corn dominating the recipe, at 74 percent of the grain bill. So for this experiment I decided to focus on just modern yellow dent corn varieties. And given that TX Whiskey works only with Texas-grown grains, we used only Texas farms for the experiment.

The experiment began with a tic-tac-toe box. Basically, it was the same setup as what I had envisioned for my wine terroir tasting. Three varieties of corn were planted on three farms: nine boxes for results. We took three different types of modern yellow dent corn from three different seed companies: D57VP51 from Dyna-Gro®, 2C797 from Mycogen Seed®, and REV25BHR26 from Terral Seed®. Then we planted them on three different farms in Texas: the Texas AgriLife Extension farm in Calhoun County, Rio Farms in Hidalgo County, and Sawyer Farms in Hill County. For good measure, we also grew a sample of the REV25BHR26 variety from Terral Seed® at the Texas AgriLife Extension farm in Hansford County. So in reality, it wasn't a true tic-tac-toe board, as we actually had ten boxes (table 8.1).

We chose these farms somewhat at random from the Texas Corn Performance Trials, which are conducted yearly by the Texas A&M AgriLife Research group. That said, we wanted farms that highlighted the different environments of Texas. All four were located within different districts of the Texas A&M AgriLife Extension Service, and each had a different soil type and local environment. Calhoun County is in the Coastal Bend, nestled along the Gulf of Mexico, rich with Livia silt loam. Rio Farms in Hidalgo County is farther west, around the border of Mexico, and built on Raymondville clay loam. Sawyer Farms is in the Central District, and it sits on Houston black clay soil. The fourth farm, in Hansford County, where

TABLE 8.1 THE TEXAS TIC-TAC-TOE

VARIETY	FARM LOCATION			
Dyna-Gro® D57VP1	Calhoun County	Hidalgo County	Hill County	
Mycogen Seed® 2C797	Calhoun County	Hidalgo County	Hill County	
Terral Seed® REV25BHR26	Calhoun County	Hidalgo County	Hill County	Hansford County

we grew the tenth sample, is in the Panhandle, on the border with Oklahoma, and it has Perryton silty clay. (Additional details on these four growing locations are listed in table 8.2.) I had no idea how the different environments or soil types would change the development of flavor compounds and their precursors in the corn. But I reasoned that these four farms, their soils, and their climate were distinct enough to represent environmental diversity. If terroir existed, then we should be able to taste it in this experiment.

The three varieties of yellow dent corn had all been bred and created relatively recently. So it's also more than likely they're relatively similar genetically. The majority of lines in modern American breeding programs are the products of a small and stratified genetic base.[2] This is because American Corn Belt varieties can be traced to a narrow range of populations within only two races: northern flint corn and southern dent corn.[3]

In other words, while seed companies might have created thousands of different corn varieties in recent decades, they all came from parents that are genetically extremely close. Think of it like blending paints. You can make loads of colors just starting with shades of red—there are hundreds of tints, from scarlet and ruby to crimson. And though they'd be different, they wouldn't be *that* different. This is essentially what corn breeders at major seed companies have been doing—breeding thousands of varieties of corn working only with shades of red. That doesn't mean the corn varieties they've created have bad flavor, but it does mean that they are all much

TABLE 8.2 CHARACTERISTICS OF THE DIFFERENT GROWING LOCATIONS

FARM OPERATION	COUNTY	EXTENSION DISTRICTS	SOIL TYPE	PLANTING DATE	HARVESTING DATE	PPOP	IRRIGATION	RW	CROP ROTATION
Texas AgriLife Extension, Port Lavaca	Calhoun	Coastal bend	Livia silt loam	2/26/2016	8/3/2016	53,987	Dryland	38 in	Grain sorghum
Rio Farms, Monte Alto	Hidalgo	South	Raymondville clay loam	2/18/2016	7/21/2016	57,027	Three times	30 in	Soybeans
Sawyer Farms	Hill	Central	Houston black clay	Middle February	Middle August	64,218	Dryland	30 in	Wheat
Texas AgriLife Extension	Hansford	Panhandle	Perryton silty clay	5/11/2016	10/11/2016	75,012	Yes	30 in	Soybeans

Note: Ppop = average plant population per hectare. Rw = average row width between rows. Sawyer Farms is the only commercial grower, with other locations being sites of the Texas A&M (TAMU) Corn Variety Testing Program.

Source: R. J. Arnold et al., "Assessing the Impact of Corn Variety and Texas Terroir on Flavor and Alcohol Yield in New-Make Bourbon Whiskey," *PloS One* 14, no. 8 (2019).

closer genetically than they would be if the breeding programs incorporated the other hundred or so recognized races of corn that exist throughout the Americas[4] or the more than twenty thousand accessions currently in the USDA-ARS National Germplasm Repository.[5] If corn breeders would only dabble in some shades of blue and yellow, the diversity of corn varieties they could create would be vast.

Working with genetically close cousins was, in part, a contrivance of our experiment. Our goal, after all, was to understand terroir within modern yellow dent corns, since those were the types used by the major bourbon distilleries, including us. By their nature, these varieties will all be genetically similar. But we wanted to explore whatever diversity was possible within modern yellow dent corn. So we chose three varieties from three different breeding programs.

To remove as many confounding factors as possible, we decided to mill, mash, ferment, and distill the corn on a laboratory scale—meaning small amounts—instead of a manufacturing scale.

For the experiment I built a small, two-liter distillation system that yielded about three ounces—100 milliliters—of new-make whiskey per batch. With help from two scientists of the Scotch Whisky Research Institute, Dr. Reginald Agu and Dr. Barry Harrison, I adapted a method meant to mimic the process of a large-scale scotch grain whisky distillery. I made just a few tweaks to more closely mirror typical bourbon production.

I had ten bags of corn. One from each variety/farm combination, and one of those varieties grown on a fourth farm. I split each bag into three separate batches. Replicating the distillation three times per bag would even out variation among the treatments and protect my results from experimental variation. If the data among the replicates from one bag were not consistent, I would know something had gone wrong with a single experiment.

Before grain is mashed or fermented, it's ground. At a distillery, we use a giant roller mill to grind thousands of pounds of grain at a time. That would be overkill, so I used a small Victoria plate mill that could grind about one pound of grain per batch. And instead of the twenty-foot stainless-steel cooker with live steam injection, I used a three-liter glass beaker and a hot plate as my mash cooker. When the mash was ready, I used a specialized flask called a Fernbach for the fermentation. And for every batch I used the same amount of TX Whiskey's proprietary yeast strain, which I first isolated from a Texas pecan nut in 2011.

After fermentation, batch distillation takes place in two steps. The first distillation is called the *stripping run*. For that I used a stainless-steel still with a fan-cooled condenser and an electric, indirect heating element. The stripping run yielded about 550 milliliters of low-proof distillate, which we call *low wines* in the industry. The second distillation is the *spirit run*. For that, I added the low wines to the spirit still, a copper alembic heated on a hotplate with a worm-coil condenser. The Scotch Whisky Research Institute has a similar lab-scale copper still manufactured specifically for them by Forsyth, the premier still manufacturer in Scotland. I bought mine off Amazon, but it still worked. In the spirit run, the 550 ml of low wines became about 100 ml of new-make whiskey—which we also call *high wines* and sometimes *white dog* in America.

Now I had thirty 100 ml samples of new-make unaged bourbon whiskey. But I needed a collaborator. To really unravel how terroir changed the chemistry and flavor of the samples, I would need to exploit some sophisticated flavor chemistry and sensory-analysis techniques. And by default, I would need to partner with someone skilled in both analytical chemistry and sensory science.

I found that collaborator in a scientist named Ale Ochoa.

9

THE CHEMISTRY OF TERROIR

EARLIER I MENTIONED that those trained in sensory science can taste a food or drink and pick out individual nuances within the overall flavor impression. To most, Coca-Cola simply tastes like, well, Coca-Cola. But sensory scientists and experienced tasters can detect the individual flavors that make up what we just think of as Coke: lemon, vanilla, cinnamon, and so on. Ale Ochoa is this type of person.

Ale was a Texas A&M Aggie from the start. She was born and raised in College Station and studied flavor chemistry and sensory science in everything from beef to coffee to whiskey. I met her through my doctoral advisor, Dr. Seth Murray. Seth is a corn breeder and quantitative geneticist, but he sometimes sits on the dissertation committees of students studying adjacent subjects.

One day in 2016, when they were both at a coffee research conference, Ale confided in Seth that, while she liked coffee, what she really wanted to work with was whiskey. At the time, I was already chugging away at my PhD under his guidance. He promptly put us in touch.

The first time I spoke with Ale over the phone, I asked her, "What got you interested in whiskey?"

"Its complexity," she said. "How such relatively simple ingredients—water, grain, yeast, and oak—can create such a wide range of flavors and styles."

The reason I remember this so well is that she used the word *simple.* That didn't exactly do the ingredients justice, I thought at first. Maybe they appear that way to someone just taking their first dive into whiskey, but discovering the complexity of the base ingredients is one of the joys of distilling.

I told her coffee might be just as complex as whiskey. There are loads of overlap between the two. This is especially true when it comes to how both beverages are analyzed at the chemical and sensory level.

Ale's graduate school advisors, Dr. Chris Kerth and Dr. Rhonda Miller, were partly responsible for developing the sensory lexicon for World Coffee Research (WCR). WCR is a nonprofit research and development organization that pursues agricultural innovation and research to secure the future of coffee against the threats of climate change and disease. They create new varieties and agronomic approaches geared at providing the farmer with the best possible yield. And while coffee farmers care about yield just as much as grain farmers, the harvested coffee beans must ultimately also make a good cup of coffee.

A sensory lexicon is a universal language for describing and quantifying flavor. It's an established tool that allows people to calibrate themselves to the same flavors and intensities and therefore find common ground and meaningful flavor data in a food or drink. It's the first step to understanding what causes a food or drink to taste, smell, and even feel the way it does. The sensory lexicon developed for the WCR is used across the world to breed and select for the most promising varieties of coffee, taking into consideration the influence of the environment in agronomic performance and flavor. And, as I saw it, it was a tool with which we could scientifically establish the flavor effects of terroir.

A panel of scientists use the sensory lexicon in a technique called *quantitative descriptive analysis* (QDA). The sensory lexicon is the training guide that allows the sensory panelists to anchor themselves to the same flavor references and intensities. They calibrate their senses using the lexicon and then identify and quantify the flavors of a particular food or beverage.

To illustrate this, let's imagine a sensory panel comprising Bugs Bunny, Daffy Duck, and the Tasmanian Devil, all there to evaluate the flavors of a new variety of carrot. Now, Bugs Bunny is obviously very familiar with the taste of carrots, their slight sweetness, earthiness, and gentle floral notes. But Daffy Duck and the Tasmanian Devil are not. And even Bugs Bunny might not know how to quantify the intensity of those flavors on a scale of, say, 0 to 15. On this scale, 0 means the flavor is not identifiable, 2 is barely detectable, 4 is identifiable, 6 is slightly intense, 8 is moderately intense, 10 is intense, 12 is very intense, and 15 is extremely intense. Mr. Warner, the sensory panel leader, realizes he must train and calibrate his panel so they are all on the same flavor page.

To do that, he mixes a sweet-aromatic compound called vanillin in water at different concentrations. The lower concentrations of vanillin in water can be used as references for the lower numbers on the intensity scale, and the higher concentrations of vanillin in water are references for the higher numbers. Mr. Warner distributes these different concentrations of vanillin in water to the panel, and they spend some time learning what the flavor of *sweet* means (in this case, the aroma of vanillin) and how their nose, tongue, and brain perceive the different intensities. The hope, then, is that when Bugs Bunny, Daffy Duck, and the Tasmanian Devil all bite into the same carrot, they will all have the same idea of what the flavor of *sweet* is and will be able to rate the intensity in the same way. If Bugs Bunny says the intensity of sweet in the carrot is 6, Daffy Duck says 7, and the Tasmanian Devil says 5, then Mr. Warner would write down 6—the averaged consensus intensity. If the panel hadn't trained on the sweet references, their answers could have varied widely, which is not useful or insightful.

Back in 2016, I knew Ale was looking for a research project to complete her master's thesis. "Why don't you switch to whiskey?" I said. On the phone, I offered her a deal: TX Whiskey would fund her master's thesis research if she would pivot to analyzing my thirty new-make bourbon samples. I explained that while my lab at the distillery was fully equipped to produce new-make bourbon, it was not equipped to run analyses on flavor chemistry or execute peer-review-caliber sensory analysis. But Ale had access to labs at Texas A&M that could.

Essentially, I wanted Ale to train an expert sensory panel from Texas A&M on flavors and references that corresponded with general flavors in new-make bourbon and then lead the panel sessions, analyzing all thirty new-make bourbons from the ten variety-farm treatments.

"One hundred percent I can do that," Ale said. "And we can also couple the data to GC-MS-O analysis. I've utilized that technique on many occasions for beef and coffee research."

I was hoping this would be a possibility. While sensory data are useful, it's even better when coupled to chemical data. Gas chromatography–mass spectrometry (GC-MS) is the most common tool for chemically analyzing flavor compounds. I had assumed this technique would be available at her lab at Texas A&M. Olfactometry (the "O" in GC-MS-O), though, was a surprise. But it was a welcome one, as it meant we could focus primarily on those compounds identified by GC-MS that actually possessed an aroma.

Gas chromatography (GC) is a technique that separates volatile compounds within a vapor mixture. The vapor is pushed with an inert or unreactive gas (often helium or nitrogen) through a column packed with a material that will bind to different compounds with varying intensities. Binding efficiency is usually based on the chemical similarity between the packing material and the specific compound—like attracts like. As the carrier gas is pushed through, those compounds not bound tightly will release from the packing material of the column—or, as we say, they *elute* from the column—more readily and make their way to the instrument detector. Those compounds that are bound tightly will take longer for the inert gas to push them through the column and to the detector. Compounds of the same type will elute together. For example, once the conditions are right that the compound isoamyl acetate (which we perceive as a banana flavor) will elute, all of the individual isoamyl acetate compounds in the column will "let go" of the packing and exit to the detector.

Mass spectrometry (MS) is a highly sensitive technique that identifies and quantifies those compounds once they leave the column. As a compound elutes from the column, it is ionized by energetic or chemical forces, and the spectrometer detects those ions. The profile of the ions detected allows us to identify the compound, and the abundance of ions lets us measure how much of that compound exists in the sample.

To make the GC-MS technique much more meaningful for flavor analysis, the scientist can incorporate olfactometry—literally "the measure of smell." As vaporized compounds elute from the column, a portion travels to the mass spectrometer, and the rest travels to the "sniff port." There a scientist waits, with his or her nose inside a heated glass nosepiece. The scientist slowly breathes in through their nose, and whenever they detect a smell, they click a button that saves the aroma event to whatever compound the mass spectrometer is detecting at that time. For example, if the scientist detects the aroma of banana at a given time, he or she can make a note of when it occurred during the analysis and then review the MS data to see which compound was detected at the time of the banana aroma event. In this situation, there's a good chance the MS would relay that the compound was isoamyl acetate or some other related compound.

So, in total, GC-MS-O stands for gas chromatography–mass spectrometry–olfactometry. The GC is how we separate out the compounds, the MS is how we detect which compounds are present, and the O is how we characterize the aroma of those compounds.

Working with Ale and her research advisors, I was confident that quantitative descriptive analysis and GC-MS-O would peel back the layers and allow us to understand which flavor compounds—and ultimately which flavors—were meaningfully influenced by terroir.

* * *

A few weeks after our phone call, I drove down from Fort Worth to College Station to hand deliver the thirty new-make bourbon samples. I also brought some of the new-make TX Bourbon we had produced the day before at the distillery. While the new-make TX Bourbon was still precious in its own right, there were thousands of gallons of it made weekly, meaning we could use it to develop and calibrate the GC-MS-O and quantitative sensory methods. Ale took the new-make TX Bourbon, placed a small volume of it inside a glass jar, covered it with nonreactive plastic, and inserted a small fiber needle that she called "SPME."

SPME, I learned, stood for *solid-phase microextraction*. SPME was the name of the technique, and the fiber needle coated with a sorbent material was the SPME fiber. Ale explained to me that while the new-make bourbon sat in the glass jar, its compounds were constantly evaporating into the headspace above the liquid and below the nonreactive plastic cover. The SPME fiber was situated right in the middle of the headspace, and the sorbent material covering the needle would adsorb the evaporated compounds.

After a few hours, Ale removed the needle and walked it to the GC-MS-O. The GC-MS-O is large, easily taking up fifteen square feet of the lab counter. She inserted the fiber into the machine, pushed a few buttons on the computer, and sat down at the glass sniff port. The evaporated compounds would elute off the SPME fiber, travel through the GC column, and then split at the fork: one half would go to the MS detector, and the other half to the sniff port. The one in the Kerth lab was unique in that it had two sniff ports, so two scientists could identify and characterize aromas. This is to minimize bias and protect against a single panelist missing a smell.

Ale and I sat at the adjacent sniff ports and placed our noses a few centimeters from the surface. "Put your nose right up to it," she told me. "You want it to rest against the glass surface."

I moved my nose just slightly until I could feel the heated surface of the glass and then tried to relax and settle in. It would take about thirty minutes for all of the individual compounds to elute from the GC column. At first, nothing. Just the background smell of the carrier gas pushing the compounds along. But then, a few minutes into the run, a faint hint of something. It was solvent and . . . sweet.

"You smell that?" Ale said. "That smells like ethyl acetate."

No surprise there. Ethyl acetate was a primary ethyl ester of all new-make whiskeys, bourbon or not. It is the ester of acetic acid and ethanol, both of which are produced under normal fermentation conditions.

Then, not a minute later, another aroma came down the line. "Ohh, I like that," Ale said. I watched her try to decide what it was. "Definitely fruity. Somewhere between apple and banana."

"Maybe it's isoamyl acetate or ethyl hexanoate," I said. These are the two compounds which smell like bananas and apples, respectively.

"Yeah, maybe. I'll look at the MS data later and see if the spectra lines up with either of those."

As we sat there and more aromas continued to come through the glass sniff ports, I found myself feeling a bit lightheaded. "Don't forget to breathe," Ale said, almost as if she could tell I was dizzy. "Most people try to concentrate so hard on their first few runs at this that they forget to breathe. It can feel like you have a head cold or even like a hangover." I moved away from the glass sniff port, looked down at my phone, and realized just how bad my head hurt.

"Oh, wow, that's chocolate!" Ale said, looking up at me excitedly, but I was no longer manning my station at the sniff port. "You forgot to breathe, didn't you?"

I left to get a drink of water and clear my head. Outside, I decided that while I thought of myself as both a scientist and a whiskey maker, I was not the type of scientist that Ale was. GC-MS-O was not my strong suit.

I went to lunch as Ale ran more samples on the GC-MS-O and fine-tuned the method. The food evened me out, and I returned to the lab a few hours later without feeling like I had spent the previous night drinking bourbon instead of just smelling it. Ale had already finished running several GC-MS-O runs and had dialed in the method. She had moved into a different lab and was now seated at the head of a table with six other scientists, the members of the sensory panel whom Ale had spent the past few weeks training to analyze whiskey, equipping them to identify and quantify more than fifty different common whiskey aromas. These aromas and their respective references were the constituents of the new-make bourbon sensory lexicon Ale had developed.

In front of each panelist were over one hundred plastic, two-ounce soufflé cups—these were the references for each of the different aromas, and most aromas had at least two different cups that spanned some range of the 0–15 intensity scale. We used everyday food and drink products as reference points for aromas, along with their own reference food score. The entire lexicon, as we would eventually publish it, is in table 9.1.

TABLE 9.1 NEW-MAKE BOURBON LEXICON

AROMA	DESCRIPTION	REFERENCE SCALE	REFERENCE PREPARATION
		AROMA FACTORS	
Alcohol	A colorless, pungent, chemical-like aromatic associated with distilled spirits or grain products	5.0: Absolut vodka (40% ABV)	Dilute 16 ml of Absolut vodka in 64 ml of distilled water. Serve 15 ml in a snifter. Cover.
		8.0: Barsol pisco (41.3% ABV)	Serve 15 ml of Barsol pisco spirit in a snifter. Cover.
		10.0: grain neutral spirit (60% ABV)	Dilute 100 g of 190 proof neutral spirit in 77.25 g of distilled water. Serve 15 ml in a snifter. Cover.
		12.0: grain neutral spirit (90% ABV)	Serve 15 ml of 190 proof neutral spirit in a snifter. Cover.
Anise	A pungent, sweet, spicy, caramelized aromatic with petroleum, medicinal, floral, and licorice scents	7.5: anise seed	Place ½ teaspoon of McCormick's anise seed in a snifter. Cover.
Banana	Aromatic characteristic of ripe bananas	10.0: banana extract	Place 1 drop of banana extract on a cotton ball. Serve in a snifter. Cover.
Barnyard	Aromatic characteristic of livestock animal housing	6.0: McCormick's ground white pepper	Place ½ teaspoon of white pepper in 1 oz. of distilled water.

Blended	The melding of individual sensory notes such that the products present a unified overall sensory experience as opposed to spikes or individual notes	3.0: Absolut vodka (40% ABV)	Dilute 16 ml of Absolut Vodka in 64 ml of distilled water. Serve 15 ml in a snifter. Cover.
		5.0: McCormick gin (40% ABV)	Serve 15 ml of McCormick gin in a snifter. Cover.
		10.0: Tanqueray gin (47.3% ABV)	Serve 15 ml Tanqueray gin in a snifter. Cover.
Brown spice complex	The sweet, brown aromatic associated with spices such as cinnamon, clove, nutmeg, and allspice	3.0: cinnamon stick	Place 2 cinnamon sticks (½ teaspoon) in a 2-oz. glass jar with screw-on lid.
		7.0: whole nutmeg and clove bud	Place 1 whole nutmeg (2 teaspoons) and 3 clove buds (¼ teaspoon) in a 2-oz. glass jar with screw-on lid.
Brown sugar	A rich, full, round, sweet aromatic impression characterized by some degree of darkness	6.0: C&H pure cane sugar, golden brown	Place 1 teaspoon brown sugar in a snifter. Cover.
Burnt	The dark brown impression of an overcooked or over-roasted product that can be sharp, bitter, and sour	4.5: benzyl disulfide	Place 0.1 gram of benzyl disulfide in a covered soufflé cup.
		8.0: puffed wheat cereal	Serve 1 tablespoon of cereal in a covered soufflé cup.

(*continued*)

TABLE 9.1 NEW-MAKE BOURBON LEXICON (CONTINUED)

AROMA	DESCRIPTION	REFERENCE SCALE	REFERENCE PREPARATION
Buttery	Aromatic associated with fresh butter, fat, sweet cream	5.0: McCormick coconut extract	Place 1 drop on a cotton ball. Serve in a snifter. Cover.
		7.0: Land O'Lakes unsalted butter	Place ½ tablespoon in a covered snifter.
Butyric	An aroma associated with butyric acid, cheesy, also sickly	6.0: butyric acid	Place 1 drop on a cotton ball. Serve in a snifter. Cover.
Caramel	A round, full-bodied, medium brown, sweet aromatic associated with cooked sugars and other carbohydrates. Does not include burnt or scorched notes.	8.0: Le Nez du Café no. 25 "caramel"	Place 1 drop of essence on a cotton ball in a soufflé cup. Cover.
Cardboard/ paper-like	The aromatic associated with cardboard or paper packaging	3.0: white napkin	Place a 2-inch napkin piece in a soufflé cup.
		7.5: cardboard	Cut a 2-inch square of cardboard. Place in a covered soufflé cup.
Coconut	The slightly sweet, nutty, somewhat woody aromatic associated with coconut	7.5: McCormick coconut extract	Place 1 drop on a cotton ball. Serve in a snifter. Cover.

Coffee	An aroma note associated with coffee	3.0: Werther's coffee	Place a single, unwrapped Werther's coffee-flavored caramel in a snifter. Cover.
		8.0: Folgers instant coffee crystals	Place ⅛ teaspoon of Folgers instant coffee crystals in a soufflé cup.
Corn	An aroma note associated with corn	5.0: canned corn	Drain and rinse canned corn and serve in a soufflé cup.
		8.0: Amoretti sweet corn essence	Place 1 drop on cotton ball and place in a soufflé cup.
Fermented/ yeasty	The pungent, sweet, slightly sour, sometimes yeasty, alcohol-like aromatic characteristic of fermented fruits or sugar or overproofed dough	5.0: Guinness Extra Stout beer	Serve 15 ml in a covered glass.
Fruity-berry	The sweet, sour, floral, sometimes heavy aromatic associated with a variety of berries such as blackberries, raspberries, blueberries, or strawberries	3.0: Captain Morgan rum	Serve 15 ml in a covered glass.
		6.0: Tropicana berry juice	Serve 115 ml in a covered glass.
		10:0: Private Selection triple berry preserves	Place 1 teaspoon in a medium snifter. Cover.
Fruity-citrus	A citric, sour, astringent, slightly sweet, peely, and somewhat floral aromatic that may include lemons, limes, grapefruits, or oranges	4.5: lemon peel and lime peel	Put 0.5 grams lemon peel and 0.5 grams lime peel in a medium snifter. Cover.
		7.5: grapefruit peel	Put 0.25 grams in a medium snifter. Cover.

(*continued*)

TABLE 9.1 NEW-MAKE BOURBON LEXICON (CONTINUED)

AROMA	DESCRIPTION	REFERENCE SCALE	REFERENCE PREPARATION
Fruity-dark	An aromatic impression of dark fruit that is sweet and slightly brown and is associated with dried plums and raisins	3.0: Sunsweet Amz!n prune juice	Mix 1 part juice with 2 parts water. This may be prepared 24 hours in advance and refrigerated. Bring to room temperature.
		4.5: Sun-Maid prunes	Chop ½ cup prunes. Add ¾ cup of water and cook in microwave on high for 2 minutes. Filter with a sieve. Place 1 tablespoon of juice in a medium snifter. Cover.
		6.0: Sun-Maid raisins	Chop ½ cup raisins. Add ¾ cup of water and cook in microwave on high for 2 minutes. Filter with a sieve. Place 1 tablespoon of juice in a medium snifter. Cover.
Fruity-other	A sweet, light, fruity, somewhat floral, sour, or green aromatic that may include apples, grapes, peaches, pears, or cherries	5.0: Le Nez du Café no. 17 "apple"	Place 1 drop on a cotton ball in large snifter. Cover.
		9.0: Effen black cherry vodka	Serve 15 ml in a covered glass.
Fishy	Aromatic associated with trimethyl-amine and old fish	7.0: canned tuna	Place 1 gram in a covered soufflé cup.
Floral	A sweet, light, slightly fragrant aromatic associated with flowers	6.0: Welch's 100% white grape juice	Mix 1 part water and 1 part juice. Place 15 ml of mixture in a snifter. Cover.
		8.0: Le Nez du Café no. 12 "coffee blossom"	Place 1 drop on a cotton ball in a snifter. Cover.

Grain complex	The light brown, dusty, musty sweet aromatic associated with grains	5.0: rice and wheat	Blend ½ cup of Rice Chex and ½ cup of Post Shredded Wheat in a food processor. Serve 1 tablespoon in a snifter. Cover.
		8.0: Georgia Moon corn whiskey	Serve 15 ml in a snifter. Cover.
Green	An aromatic characteristic of fresh, plant-based material. Attributes may include leafy, viney, unripe, grassy, and peapod.	9.0: parsley water	Rinse and chop 25 grams of fresh parsley. Add 300 ml of water. Let sit for 15 minutes. Filter out the parsley. Serve 1 tablespoon of the water in a snifter. Cover.
Hay-like	The lightly sweet, dry, dusty aromatic with slight green character associated with dry grasses	7.5: McCormick parsley flakes	Place 1 teaspoon in a medium snifter. Cover.
Herb-like	The aromatic commonly associated with green herbs that may be characterized as sweet, slightly pungent, and slightly bitter. May or may not include green or brown notes.	3.0: McCormick bay leaves, ground thyme, basil leaves	Mix together 0.5 grams of each herb. Break the bay leaves into smaller pieces with your hands first, and then grind all the herbs together using a mortar and pestle. Add 100 ml of water. Mix well. Put 5 ml of herb water in a medium snifter and add 200 ml of water. Serve 1 oz. in a soufflé cup.
		10.0: McCormick bay leaves, ground thyme, basil leaves	Mix together 0.5 grams of each herb. Break the bay leaves into smaller pieces with your hands first, and then grind all the herbs together using a mortar and pestle. Add 100 ml of water. Mix well. Serve 1 oz. in a soufflé cup.

(*continued*)

TABLE 9.1 NEW-MAKE BOURBON LEXICON (CONTINUED)

AROMA	DESCRIPTION	REFERENCE SCALE	REFERENCE PREPARATION
Honey	Sweet, light brown, slightly spicy aromatic associated with honey	6.0: Busy Bee pure clover honey	Dissolve 1 tablespoon honey in 250 ml of distilled water. Serve 15 ml in snifter. Cover.
Lactic acid	A sour aroma note associated with lactic acid	5.0: buttermilk	Serve 1 oz. buttermilk in a soufflé cup.
		8.0: sauerkraut	Serve 5 g of sauerkraut in a soufflé cup.
Leather	An aromatic associated with tanned animal hides	3.0: leather shoe lace	Place a three-inch length of leather shoelace in a covered snifter.
		10.0: Hazels Gifts leather essence	Place two drops on a cotton ball in a covered snifter.
Malt	The light brown, dusty, musty, sweet, sour, and/or slightly fermented aromatic associated with grains	3.5: Post Grape-Nut cereal	Serve in a covered snifter.
		6.0: Carnation malted milk	Place ½ teaspoon in a covered snifter.
Medicinal	A clean, sterile aromatic characteristic of antiseptic-like products such as Band-Aids, alcohol, and iodine	6.0: Le Nez du Café no. 35 "medicinal"	Place 1 drop on a cotton ball in a soufflé cup.
		8.0: Tanqueray gin	Serve 15 ml in covered glass.
		12.0: iodine	Serve 1:1 iodine and distilled water solution in a covered glass (50 ml iodine tincture, 50 ml distilled water).

Mint	An aromatic that is sweet, green, and menthol	4.0 Absolut vodka/mint gum	Place 3 sticks of mint gum in 150 ml of Absolut vodka and let steep for 30 minutes. Serve 15 ml in a covered glass.
		8.0: Listerine	Serve in a covered snifter.
Molasses	An aromatic associated with molasses; has a sharp, slight sulfur or caramelized character	6.5: blackstrap molasses	Mix 2 teaspoons in 250 ml of water. Serve ¼ cup in a mason jar. Cover.
Musty/dusty	The aromatic associated with dry, closed-air spaces such as attics and closets. May have elements of dry, musty, paper, dry soil, or grain.	5.0: Kretschmer wheat germ	Serve 1 tablespoon in a medium snifter. Cover.
		10.0: 2,3,4-trimethoxybenzaldehyde	Place 0.1 gram in a medium snifter. Cover.
Musty/earthy	The somewhat sweet, heavy aromatic associated with decaying vegetation and damp, black soil	3.0: mushrooms	Place two washed ½-inch cubes in a covered snifter.
		9.0: Miracle-Gro potting soil	Fill a 20-oz. glass jar half full with potting soil and seal tightly with screw-on lid.
		12.0: Le Nez du Café no. 1 "earthy"	Place 1 drop on a cotton ball in a large snifter. Cover.
Nutty	A slightly sweet, brown, woody, oily, musty, astringent, and bitter aromatic commonly associated with nuts, seeds, beans, and grains	7.5: Le Nez du Café no. 29 "roasted hazelnut"	Place 1 drop on a cotton ball in a covered glass.
		9.0: almond/walnut puree	Puree the almonds and walnuts separately in blenders for 45 seconds on high speed. Combine equal amounts of the chopped nuts. Serve in a covered glass.

(*continued*)

TABLE 9.1 NEW-MAKE BOURBON LEXICON (CONTINUED)

AROMA	DESCRIPTION	REFERENCE SCALE	REFERENCE PREPARATION
Oily	An overall flavor term for the aroma and flavor notes reminiscent of vegetable oil or mineral oil products	9.0: vegetable oil	Serve in a covered glass.
Overall sweet/ sweet aromatics	The perception of a combination of sweet substances and aromatics	3.0: vanillin	Mix 0.5 g in 250 ml of water and serve in a covered snifter.
		5.0: vanillin	Mix 2 g in 250 ml of water and serve in a covered snifter.
Overall sour/sour aromatics	An aromatic associated with the impression of a sour product	2.0: Bush's pinto beans, canned	Drain and rinse with distilled water. Place 1 tbsp. in a covered snifter.
		5.0: buttermilk	Serve 1 oz. in a covered glass.
Pepper	The spicy, pungent, musty, and woody aromatic characteristic of ground black pepper	13.0: McCormick ground black pepper	Place ½ teaspoon in a medium snifter. Cover.
Rancid	Aromatic associated with oxidized fats and oils	5.0: vegetable oil (oxidized/rancid)	Keep oil in an open container or warm storage place for 1 week. Place 1 oz. in a covered glass.
Roast	Dark brown impression characteristic of products cooked to a high temperature by dry heat. Does not include bitter or burnt notes.	6.0: Le Nez du Café no. 34 "roasted coffee"	Place 1 drop on cotton ball. Place in a covered glass.

Smoky	An acute, pungent aromatic that is a product of the combustion of wood, leaves, or a non-natural product	6.0: Diamond smoked almonds	Place 5 almonds in a covered snifter.
Soapy	An aroma associated with unscented soap	6.5: Ivory soap flakes	Place 0.5 g in 100 ml of room temperature water. Serve in a large, covered snifter.
Solvent-like	General term used to describe many classes of solvents, such as acetone, turpentine, chemical solvents, etc.	5.0: acetone solution	Dilute 10 ml in 100 ml distilled water until dissolved, and serve in a 2 oz. soufflé cup. Cover.
		8.0: lighter fluid solution	Dilute 10 ml in 100 ml distilled water until dissolved, and serve in a 2 oz. soufflé cup. Cover.
Stale	The aromatic characterized by a lack of freshness	4.5 Mama Mary's Gourmet Original pizza crust	Cut a 2-inch square of crust and serve in a soufflé cup. Cover.
Sulfur	Aromatic associated with hydrogen sulfide, rotten eggs	3.0: Bush's pinto beans	Drain and rinse the beans. Serve 1 tbsp. in a covered glass.
		11.0: dimethyl trisulfide	Dilute 1 ml in 100 ml distilled water until dissolved, and serve in a 2 oz. soufflé cup. Cover.
		15.0: dimethyl trisulfide	Place 1 drop on a cotton ball in a large snifter. Cover.

(*continued*)

TABLE 9.1 NEW-MAKE BOURBON LEXICON (CONTINUED)

AROMA	DESCRIPTION	REFERENCE SCALE	REFERENCE PREPARATION
Tobacco	The brown, slightly sweet, slightly pungent, fruity, floral, spicy aromatic associated with cured tobacco	5.0 Le Nez du Café no. 33 "pipe tobacco"	Place 1 drop on a cotton ball in a large snifter. Cover.
		7.0 Marlboro cigarettes, Southern Cut	Break cigarette and place 0.1 grams tobacco in a medium snifter. Cover.
Vanilla	A woody, slightly chemical aromatic associated with vanilla bean, which may include brown, beany, floral, and spicy notes	2.5: Le Nez du Café no. 10 "vanilla"	Place 1 drop on a cotton ball in a snifter. Cover.
		5.5: Spice Islands bourbon vanilla bean	Place 0.5 gram chopped vanilla beans in a snifter. Cover.
Vinegar	A sour, astringent, slightly pungent aromatic associated with vinegar or acetic acid	2.0: 0.5% acetic acid solution	Dilute 5 ml distilled white vinegar in 1,000 ml distilled water. Serve in a soufflé cup. Cover.
		3.0: 2.0% acetic acid solution	Dilute 20 ml distilled white vinegar in 1,000 ml distilled water. Serve in a soufflé cup. Cover.
Woody	The sweet, brown, musty, dark aromatic associated with a bark of a tree	4.0: Diamond shelled walnuts	Serve 1 tablespoon of chopped walnuts in a snifter. Cover.
		7.5: popsicle sticks	Break popsicle sticks in two and place in a snifter. Cover.

	NASAL FEELING FACTORS		
Nose cooling	The chemical feeling factor or sensation of cooling in the nasal passages when sniffing	6.0: Tanqueray gin	Serve 15 ml in covered glass.
		8.0: Listerine solution	Mix 1:1 dilution of Listerine and distilled water; serve in soufflé cups.
		12.0: Listerine	Serve 1 oz. in a covered glass.
Nose drying	The chemical feeling factor or sensation of drying in the nasal passages when sniffing	4.0: Barrelstone Cellars merlot, 2013	Serve 15 ml in a covered glass.
		6.0: grain neutral spirit (60% ABV)	Add 100 g of grain neutral spirit to 77.25 g distilled water; serve in a covered glass.
		8.0: unscented hand sanitizer	Serve 1 oz. in a covered glass.
Nose warming	Chemical feeling factor described as warmth or a burning sensation in the nasal passage occurring when sniffing	3.0: Barrelstone Cellars merlot, 2013	Serve 15 ml in a covered glass.
		7.0: TX blended whiskey (41% ABV)	Serve 15 ml in a covered glass.

(*continued*)

TABLE 9.1 NEW-MAKE BOURBON LEXICON (CONTINUED)

AROMA	DESCRIPTION	REFERENCE SCALE	REFERENCE PREPARATION
		9.0: grain neutral spirit (60% ABV)	Add 50 g of grain neutral spirit to 79.8 grams distilled water; serve in a covered glass.
		12.0: grain neutral spirit (60% ABV)	Add 100 g of grain neutral spirit to 77.25 grams distilled water; serve in a covered glass.
Prickle/pungent	A feeling factor that can range from tingling or irritating; a sharp, physically penetrating sensation of the nasal cavity	5.0: horseradish sauce	Serve 1/8 teaspoon in a covered glass.
		7.0: Captain Morgan rum	Serve 15 ml in a covered snifter.
		9.0: McCormick ground black pepper	Serve ½ teaspoon in a covered glass.
		10.0: horseradish sauce	Mix 5 grams horseradish sauce in 30 ml distilled water; serve 1 oz. in soufflé cup.

To analyze one new-make bourbon sample took them no longer than fifteen minutes. In assessing each aroma, they would smell the corresponding references to calibrate themselves to the lexicon. Now, each of the more than fifty aromas was usually not identified in every sample, meaning that there might be a whole string of aromas they skipped. But I was still shocked at how quickly they could identify and quantify the whiskey smells.

You may be wondering, "Why just smell?" Why didn't the panelist also taste the new-make bourbon? The short answer is so they didn't get drunk. The long answer is that with distilled spirits, aroma analysis accounts for nearly the entire flavor assessment. You can smell almost everything you can taste. And the high ethanol content in distilled spirits has a suppressive effect on sensations. Compared to beer and wine, the impact of taste is relatively low in developing a whiskey's flavor. The creation of flavor really comes from orthonasal and retronasal olfaction. That's why orthonasal olfaction (sniffing) is the main form of sensory evaluation used in the whiskey industry.[1]

* * *

Over the following weeks, Ale repeated the chemical and sensory analyses using the thirty new-make bourbon samples from each variety-farm treatment. This is what we found.

The GC-MS-O identified and quantified sixty-eight different flavor compounds that registered an aroma event—meaning that Ale detected an aroma at the same time the MS detected a mass signal. Of the sixty-eight compounds, thirty-six of these showed *substantial variation* in their presence and concentration among different samples. This means that variation must have been caused by the corn variety, the farm, or the interaction of the two.

In plant breeding nomenclature, the variety would be called the *Genetic Effect* (G), the farm is the *Environment Effect* (E), and the interaction is called the *Genetic × Environment Interaction Effect* (GxE). Back in 2016 and 2017, as this research was being conducted and drafted, I still believed terroir was simply a synonym for E. Terroir was the farm that grew the plant.

But now I believe terroir should—at a minimum—also consider the interaction of the variety and the farm (GxE), how varieties express flavor in and through different environments.

For us, "substantial variation" meant at least a 20 percent difference in the presence and concentration of a flavor compound. And this is what was extremely promising: of the thirty-six compounds with substantial variation between the samples, eighteen were esters, five were aldehydes, and four were ketones. These are all very important classes of flavor compounds. And while 20 percent variation might not seem like a lot, it is enough for there to be statistically significant variations among the treatments. The thirty-six flavor compounds meaningfully affected by terroir, as well as their respective aromas, are listed in table 9.2.

We immediately saw that esters were the dominant class that terroir seemed to influence. Why is this interesting? For one, esters are arguably the most important flavor compound class in wine and whiskey. The analytical chemist Vicente Ferreira and his team found that of the eighteen flavor compounds present in all wines that are responsible for wine's "global odor,"[2] eight of them were esters. But the real reason I found this interesting is that while esters are synthesized by grapes and grains, they are present at such low concentrations in the original fruit or grain that their effect on a wine or whiskey's flavor will be insignificant. That means the esters that are so important for flavor in wine and whiskey are produced by yeast (and indigenous bacteria) during fermentation. This told me something important: I had to focus not just on flavor compounds that derive directly from grain but also those compounds in grain that serve as fermentation precursors.

This may seem surprising, but is it really? Remember, we are feeding these grains to the yeast as food in the mash. Many of the constituents of grain serve as the substrate building blocks of fermentation. Metabolically, these blocks are either broken down into simpler forms (catabolism) or built up into more complex forms (anabolism). They are the blocks from which yeast will disassemble and assemble every facet of their metabolism, including the byproducts excreted into the fermentation medium that become potential flavor compounds.

TABLE 9.2 THE THIRTY-SIX FLAVOR COMPOUNDS IDENTIFIED IN NEW-MAKE BOURBON THAT WERE SIGNIFICANTLY IMPACTED BY TERROIR

CLASSIFICATION	FLAVOR COMPOUND	AROMA
Acetal	1,1-diethoxyethane	Sweet, green, ethereal
Aldehyde	2-heptenal	Apple, fatty
Aldehyde	2-nonenal	Stale bread, cardboard
Aldehyde	2-octenal	Green, fatty
Aldehyde	2,4-decadienal	Meaty, fatty
Aldehyde	Nonanal	Soapy, fatty
Aromatic hydrocarbon	Napthalene	Mothball
Benzene	4-vinylanisole	Green
Benzene	Styrene	Sweet, phenolic, plastic
Ester	2-methylbutyl decanoate	Fruity
Ester	Ethyl 2-octenoate	Fruity
Ester	Ethyl 4-hexenoate	Fruity
Ester	Ethyl 2-nonenoate	Fruity
Ester	Ethyl acetate	Fruity, ethereal
Ester	Ethyl decanoate	Apple, waxy
Ester	Ethyl laurate	Floral, waxy
Ester	Ethyl hept-2-enoate	Fruity
Ester	Ethyl heptanoate	Fruity, pineapple, grape
Ester	Ethyl hexanoate	Fruity, apple
Ester	Ethyl nonanoate	Fruity, nutty
Ester	Ethyl octanoate	Fruity, floral, banana, pineapple
Ester	Ethyl sorbate	Fruity, ethereal
Ester	Ethyl trans-4-decenoate	Fruity, citrus
Ester	Ethyl undecanoate	Soapy, waxy
Ester	Isoamyl acetate	Banana
Ester	Isoamyl octanoate	Fruity, coconut, pineapple
Ester	Isopentyl hexanoate	Fruity
Furan	2-methyl-5-isopropenylfuran	Nutty, roasted
Furan	2-pentylfuran	Fruity, green
Fusel alcohol	Phenethyl alcohol	Floral, rose
Ketone	2-nonanone	Floral, fruity, cheese
Ketone	2-tridecanone	Coconut, cheese, meaty
Ketone	2-undecanone	Fruity, fatty, pineapple
Ketone	Acetophenone	Almond, cherry, orange
Organooxygen	Diethyl ether	Ethereal, sweet
Terpene	Cedr-8-ene	Woody, tobacco

Source: R. J. Arnold et al., "Assessing the Impact of Corn Variety and Texas Terroir on Flavor and Alcohol Yield in New-Make Bourbon Whiskey," *PloS One* 14, no. 8 (2019).

Let's take the results a step further. How would a distiller go about using this information to target specific flavors in a whiskey?

Take, for example, the compound 2-undecanone. It's a ketone that smells like a fruity, fatty pineapple. We found that for this flavor compound, the corn variety accounted for 23 percent of the variation and the interaction between the corn variety and the farm accounted for 63 percent. Specifically, the data showed that *only* the new-make bourbon produced from corn grown at Sawyer Farms contained 2-undecanone. And the three varieties from Sawyer Farms all expressed distinct differences in its presence and concentration. The Dyna-Gro new-make didn't contain any 2-undecanone. The Terral Seed new-make had an average to low level. But the new-make produced from the Mycogen Seed® variety had consistently high concentrations.

So if I was targeting this fruity pineapple flavor in my bourbon—a flavor typically more common in the delicate Speyside scotch single malts than in bourbon—I'd ask John Sawyer to grow the Mycogen Seed variety.

On the surface, this all seems straightforward enough. But of course it isn't this simple. Flavor is complex. Just because a flavor compound is present in a food or beverage doesn't mean that we'll actually taste it. It might be there in subthreshold quantities, or it might be masked by the synergistic effect of other flavors. While 2-undecanone appeared to be a prime candidate for terroir-driven flavor variation, how did it—or any other of the thirty-six flavor compounds detected—actually change the flavor of the samples? We needed to identify correlations between the flavor compounds and the aromas. In other words, as certain flavor compounds increased or decreased in concentration, would specific aromas increase or decrease in kind? Only these correlations would count as evidence for terroir.

From the more than fifty aromas in the sensory lexicon, the human panel identified thirteen aromas that were meaningfully influenced by terroir: alcohol, sweet, sour, grainy, corn, malt, woody, earthy, molasses, anise, lactic acid, stale, and pungent. The corn aroma had the strongest variation: it swung by 40 percent depending on the interaction of the variety and farm. Interestingly, the same new-make bourbon made from the Mycogen Seed variety grown at Sawyer Farms that contained high levels of 2-undecanone also had the highest corn aroma values.

This was promising. Thirteen aromas were significantly influenced by terroir. But meaningful correlations between individual aromas and compounds were hard to pin down. The quantified aroma concentrations were often so close that we were hard-pressed to discover differences.

So I decided to take a different approach. I decided not to consider each aroma individually but instead to group each into one of two categories—"good" and "bad." (What can I say? I didn't see the need to get too fancy with the language.)

I grouped aromas like corn, sweet, malt, banana, caramel, citrus fruit, dark fruit, floral, honey, molasses, and nutty into the good category. The aromas like sour, musty, lactic acid, grainy, burnt, stale, and sulfur I filed under bad. I summed the quantified values for all good aromas and did the same for all bad aromas. This gave me two different summed aroma intensity values—good and bad—for each of the thirty batches. The idea was simple—if a batch had a high good value, then it had overall desirable flavors. If it had a high bad value, then it had a higher concentration of bad flavors.

Of the sixty-eight different flavor compounds identified by GC-MS-O, we found eight total compounds that correlated with either the good or bad aroma values. Specifically, seven possessed significant correlations with the good aroma value, and one with the bad aroma value. It just so happened that seven of these eight flavor compounds were also included in the group of thirty-six that were affected by terroir. These compounds are listed in table 9.3.

Four of the seven that correlated with the good aroma value were esters—isoamyl acetate, ethyl nonanoate, ethyl octanoate, and ethyl 4-hexenoate—which all impart fruity and floral flavors. The fifth was nonanal, an aldehyde that tastes soapy and fatty. In balanced amounts, these flavors are good in whiskey. The sixth was styrene, which humans taste as a mix of sweet, phenolic, and even plastic flavors. As with the soapy and fatty flavor of aldehyde, a balance of phenolic notes can bring complexity to a whiskey. Some whiskey styles even pursue high concentrations of

TABLE 9.3 FLAVOR COMPOUNDS SHOWN TO SIGNIFICANTLY CORRELATE WITH "GOOD" AND "BAD" AROMAS IN NEW-MAKE BOURBON WHISKEY

CLASSIFICATION	FLAVOR COMPOUND	AROMA	CATEGORY	AFFECTED BY TERROIR
Aldehyde	Nonanal	Soapy, fatty	Good	Yes
Aldehyde	Acetaldehyde	Green apple	Good (negative correlation)	No*
Aldehyde	2-nonenal	Stale bread, cardboard	Bad	Yes
Benzene	Styrene	Sweet, phenolic, plastic	Good	Yes
Ester	Isoamyl acetate	Banana	Good	Yes
Ester	Ethyl nonanoate	Fruity, nutty	Good	Yes
Ester	Ethyl octanoate	Fruity, floral, banana, pineapple	Good	Yes
Ester	Ethyl 4-hexenoate	Fruity	Good	Yes

*Acetaldehyde was not meaningfully impacted by terroir, although it is worth noting that 15 percent of the variation in acetaldehyde concentration was attributable to the farm or the interaction of the variety and farm.

Source: R. J. Arnold et al., "Assessing the Impact of Corn Variety and Texas Terroir on Flavor and Alcohol Yield in New-Make Bourbon Whiskey," *PloS One* 14, no. 8 (2019).

phenolic flavors. Some of the most famous Islay scotch whiskies use peated malt, which imparts those characteristic smoky, rubbery, medicinal, and phenolic flavors. These six compounds showed positive correlation, meaning that when their concentrations increased, so did the good aroma value.

The seventh compound was unique: it showed negative correlation, so that when its concentration increased, the concentration of the good aroma value decreased. This was acetaldehyde, a compound that imparts flavors of astringent unripe green apples. While a balanced level of acetaldehyde can deliver desirable notes of soft, sour green apple (which is characteristic of many well-aged whiskeys), high levels deliver an overly astringent, nail polish taste.

Only one flavor compound from the original thirty-six identified showed a significant correlation to the bad aroma value—2-nonenal. This compound is a known troublemaker. It smells like stale bread and cardboard. When concentrations of 2-nonenal increased, so did the concentrations of the bad aroma values. When I try to detect faults in any new-make whiskey—not just bourbon—I always look for that smell of staleness and cardboard.

Now, this approach didn't allow us to unravel which specific flavor compounds were responsible for specific flavors in the new-make bourbons. Our approach was not sophisticated enough for that. But it did let us determine—at least to some extent—which flavor compounds important for driving some aspect of flavor in the assessed new-make bourbons were also meaningfully changed by terroir. Further, I was encouraged by the fact that many of the thirty-six flavor compounds we identified have also been shown to be not just important for flavor in wine—especially isoamyl acetate, one of the eighteen global wine odor compounds—but also influenced by terroir.

* * *

I still felt that the chemical roadmap wasn't complete. We had investigated only new-make bourbon, and we couldn't say definitively which of the thirty-six terroir-affected flavor compounds actually played a role in flavor. Hundreds of compounds can be identified in a whiskey (and many other foods and beverages), but that doesn't mean each one changes the taste.

But there are scientific techniques beyond the ones we used that can peel back the chemical layers of flavor further. I discovered one such technique for the first time while reading through the whiskey science literature, in two papers written by Peter Schieberle and his team at the Technical University of Munich and published in the *Journal of Agricultural and Food*

Chemistry. These papers reported for the first time which compounds in bourbon were responsible for flavor. The technique is called *sensomics*.

What I realized as I reviewed and referenced Schieberle's work was that sensomics might be more than just a roadmap to the flavor of whiskey or a set of data I could compare to mine. The whiskey sensomics research conducted so far has indicated that there are between thirty and sixty different compounds that contribute meaningfully to the flavor of whiskey. This list of compounds might round out the homology comparison with wine that I had originally envisioned. If any of these compounds on the whiskey sensomics list were also found in wine, and if any of those shared compounds came from grapes in wine, and if any of *those* were provably influenced by terroir in wine, then that might give me not just a chemical basis for flavor but a global roadmap to the terroir of whiskey.

10

THE ROADMAP

SENSOMICS IS A cutting-edge technique that connects analytical chemistry to sensory science to uncover which compounds in a food or beverage are actually responsible for its flavor profile. It can even assess the specific effect that a single compound (or group of compounds) has on that flavor profile. Peter Schieberle and his team used sensomics to unravel the most important flavor compounds in one particular bourbon whiskey.

In 2008, Schieberle and his team experimented with a single brand of Kentucky straight bourbon that they purchased in a supermarket in Germany. We don't know the exact brand, but any Kentucky bourbon available at this time would have used modern yellow dent corn varieties sourced from one or multiple commodity grain elevators. Up until 2014, about 60 percent of the corn used by Kentucky distilleries came from grain elevators in Indiana.[1]

The researchers took a 25 ml sample of the supermarket-bought bourbon and mixed it with diethyl ether, an effective solvent for extracting organic compounds, both volatile and nonvolatile. They used high-vacuum distillation to separate the nonvolatile and volatile compounds. The volatile fraction was condensed and then added to a liquid mixture of 40 percent ethanol in water. The scientists agreed that it smelled just like the original whiskey, and they concluded that their method had successfully captured the flavor compounds responsible for the aroma of the bourbon.

Next they separated the volatile fraction distillate into its constituent flavor compounds using gas chromatography–olfactometry. The team detected fifty unique aroma events, meaning fifty instances when a scientist at the sniffing point smelled something.

These results alone would be meaningful enough to continue on to compound identification. But sensomics goes one step further. After the volatile fraction distillate had been assessed in its original, concentrated form, they diluted it in a stepwise fashion—1:2, 1:4, 1:8, 1:16, 1:32, 1:64, 1:128, 1:256, 1:512, 1:1,024, 1:2,048, and 1:4,096. Then they assessed each diluted fraction again with GC-O. This let the scientists assign a flavor dilution (FD) factor to each aroma event, that is, essentially, to each flavor compound. The FD factor value represents the last dilution in which the aroma for a given flavor compound was still detectable. The higher the FD factor, the more potent or concentrated the flavor compound. This specific technique of sensomics is known as aroma extract dilution analysis (AEDA).

Let me explain with an analogy. Imagine that you have five separate batches of equally measured coffee grounds. You also have an imaginary coffee machine that has no limit to how much water it can hold and brew. For that first batch of coffee grounds, you brew eight cups. The flavor is intense and bold. For the second batch, you double the amount of water and brew the same amount of coffee grounds into sixteen cups. The coffee is still flavorful, but it's weaker than the first batch. For the third batch, you double the amount of water again to thirty-two cups. Now the coffee is significantly weaker, more like coffee-flavored water. The fourth batch uses sixty-four cups, and though it is dull and bland, you can detect just a trace of the hint of coffee. The fifth uses 128 cups, and you can't taste any coffee at all. In this example of AEDA, the FD factor for the coffee would be sixty-four, as that was the last concentration at which the taste of coffee was detectable.

Using AEDA, the research team found that the two most potent flavor compounds in their selected bourbon smelled of cooked apple and coconut. Other potent flavor compounds smelled of vanilla, flowers, clove, peach, and malt, and many flavor compounds had similar aromas. In total, they detected fifty flavor compounds, all with FD factors that ranged from 32 (a low FD factor) to 4,096 (the highest FD factor possible in their study). At this point, though, the scientists did not know which flavor compounds were responsible for the aroma events. To determine this, they would need to confirm the chemical structures of those fifty flavor compounds and quantify their concentration. The researchers used mass spectrometry for compound confirmation and quantification.

While FD factors are informative, they do not technically reveal exactly how or to what extent a particular flavor compound changes the flavor of a specific food or drink. To make up for the limitations of FD factors, Schieberle's team determined something called *odor activity values* (OAVs). OAVs measure the importance of a flavor compound to the overall aroma of a food or beverage. To calculate the value, they first measured the odor threshold of each flavor compound in a 6:4 water/ethanol matrix. This odor threshold represents the lowest concentration at which a flavor compound can still be detected by a human nose. They then calculated the OAV by dividing the concentration of the flavor compound in the bourbon by its odor threshold. If the OAV is lower than one, that meant the concentration of the flavor compound is below its threshold. On the other hand, if the OAV is greater than one, that meant the concentration of the flavor compound is above its threshold. The higher the OAV for a particular flavor compound, the more likely that it is an important contributor to flavor.

Let's return to the coffee analogy. Pretend you have a cup of coffee that you have analyzed with mass spectrometry. The data revealed ten flavor compounds present in different concentrations. Since you know the identity of each of the ten flavor compounds, you order pure samples of each from your favorite chemical supply company. Then you spike each pure flavor compound into water at different concentrations, from high to low. Within that range of concentrations, there will be a point somewhere at which the aroma of the flavor compound is still slightly detectable but if it was any lower in concentration it would be undetectable. That last concentration where the aroma is still detectable is called its *odor threshold.*

So you measure the odor thresholds for each of the ten flavor compounds. Then you compare this threshold to their concentrations in the coffee. Let's just consider one flavor compound as an example—vanillin. This compound is just as important in coffee as it is in whiskey. You measured the odor threshold of vanillin to be 0.02 milligrams per liter. The mass spec data showed that the concentration of vanillin in the coffee was 4.8 milligrams per liter. With these two values, you can calculate that the OAV for vanillin in this cup of coffee is 240 (4.8 divided by 0.02). And given that 240 is well above an OAV of 1, you conclude that vanillin is key to developing the flavor of the coffee.

On the other hand, if the mass spec data showed that the concentration of vanillin was only 0.005 milligrams per liter, then this would be an OAV of 0.25 (0.005 divided by 0.02), and you would conclude that the compound was present at too low a concentration to matter for the coffee's flavor.

Schieberle and his team found twenty-five flavor compounds in the bourbon (not including ethanol itself) that clearly exceeded their odor thresholds in the 6:4 water/ethanol matrix, with OAVs ranging from 2 to 138. With this information, the researchers spiked those twenty-five flavor compounds into one 6:4 water/ethanol matrix at the concentration measured in the bourbon sample, and they called this mixture the *aroma recombinate.* And it smelled like their original bourbon sample. This meant that this suite of twenty-five flavor compounds was sufficient to mimic the aroma of the bourbon and that not all fifty flavor compounds identified in their first paper were necessary for that mimicry. Then they built a series of aroma recombinates that *omitted* particular flavor compounds, so they could pinpoint their specific contribution to flavor.

Ale and I had taken an untargeted approach. We considered any flavor compound we detected as potentially important for flavor. We didn't calculate FDs or OAVs. If we did, then we likely would have found that some of those flavor compounds significantly affected by terroir (that group of thirty-six in table 9.2) were actually present at concentrations too low in the whiskeys to contribute to flavor.

The dozens of flavor compounds identified by Schieberle and his team (table 10.1) can be grouped into eight classes of flavor compounds: aldehydes, esters, fusel (higher) alcohols, ketones, lactones, norisoprenoids (a class of terpenes), volatile phenols, and a miscellaneous bucket labeled "other," which contained classes of flavor compounds that appeared only once. Every compound in these classes originated either from grain, fermentation, oak barrel maturation, or some combination of the three. The ones that potentially mattered for terroir were those derived from grain and those produced during fermentation from grain-derived precursors. Grain genetics and growing environment can affect the presence and concentration of compounds from both categories.

I decided to chart the ingredient and process origins of these compounds, their individual aromas, and whether they were present in wine (or beer). And then, most importantly, whether any of those shared grain/grape-derived flavor compounds were reportedly influenced by terroir in wine (or beer). I placed all the data into table 10.1, the first chemical roadmap of terroir in bourbon. (The sources of the data for this table can be found in appendix 2, "Key to the Roadmap.")

Looking over Schieberle's results, the first thing that jumped out to me were the norisoprenoid terpenes, and specifically β-damascenone (*cooked apple* aroma, or even sometimes described simply as *whiskey*). This flavor compound had the highest FD factor, the fourth highest OAV, and was included in the aroma recombinate. These were incredibly meaningful discoveries. β-damascenone had been identified by Vicente Ferriera and his research team as the only grape-derived compound (the other seventeen being fermentation derived) that contributes to the "global odor" of wine.[2] Further, I found research showing again and again that the concentration of β-damascenone in wine is influenced by grape variety and vineyard environment. For example, β-damascenone is reported to have a higher FD factor in grenache wine than in merlot and cabernet sauvignon,[3] and its concentration in cabernet sauvignon wine is increased when the vines are grown with cover crops.[4]

β-damascenone was not the only norisoprenoid terpene detected by Schieberle's team. α-damascone (*cooked apple*) and β-ionone (*violets*) also contained high FD factors, although they had OAVs below 1 and therefore were not included in the aroma recombinate. Norisoprenoids in grapes and grains are derived from carotenoid molecules, especially carotenes and luteins.

TABLE 10.1 CHARTING THE CHEMICAL ROADMAP OF TERROIR IN BOURBON

CLASSIFICATION	IMPORTANT FLAVOR COMPOUND IN BOURBON[1]	AROMA	ORIGIN(S)	REPORTED IN WINE OR BEER	INFLUENCED BY TERROIR IN WINE OR BEER
Acetal	1,1-diethoxyethane†	Sweet, green, ethereal	Fermentation, maturation	Wine / beer	Wine[2]
Aldehyde	2,4-nonadienal	Fatty, melon	Grain	Beer	—
	2,6-nonadienal†	Cucumber		Wine / beer	Wine[3] / beer[4]
	2-decenal	Orange, floral		Beer	—
	2-heptenal	Apple, fatty		Wine / beer	Wine[5] / beer[6]
	2-nonenal†	Stale bread, cardboard		Wine / beer	Wine[7] / beer[8]
	2-methylbutanal	Chocolate		Wine / beer	Wine[9] / beer[10]
	2,4-decadienal†	Meaty, fatty		Wine / beer	Wine[11]
	Isobutyraldehyde†	Grainy		Wine / beer	Wine[12] / beer[13]
	Isovaleraldehyde†	Chocolate		Wine / beer	Wine[14] / beer[15]
	Nonanal	Soapy, fatty		Wine / beer	Wine[16]
	Acetaldehyde*	Green apple	Fermentation, maturation	Wine / beer	Wine[17] / beer[18]
Ester	2-phenethyl propionate	Floral, rose, sweet	Fermentation	Wine / beer	—
	Ethyl 2-methylbutyrate*†	Apple		Wine / beer	Wine[19]
	Ethyl 2-phenylacetate	Cocoa, honey, floral		Wine / beer	Wine[20]
	Ethyl butyrate*†	Pineapple		Wine / beer	Wine[21]
	Ethyl cinnamate†	Cinnamon		Wine / beer	Wine[22]

	Ethyl hexanoate*†	Fruity, apple		Wine / beer	Wine[23]
	Ethyl isobutyrate*†	Citrus, strawberry		Wine / beer	Wine[24]
	Ethyl isovalerate*†	Apple, pineapple		Wine / beer	Wine[25]
	Ethyl octanoate*†	Fruity, floral, banana, pineapple		Wine / beer	Wine[26]
	Ethyl pentanoate	Apple, pineapple		Wine / beer	Wine[27]
	Ethyl propanoate	Grape		Wine / beer	Wine[28]
	Isoamyl acetate*†	Banana		Wine / beer	Wine[29]
	Phenylethyl acetate†	Honey		Wine / beer	Wine[30] / beer[31]
	Ethyl acetate*†	Fruity, ethereal	Fermentation, maturation	Wine / beer	Wine[32] / beer[33]
Fusel alcohol	Isoamyl alcohol*†	Banana	Fermentation	Wine / beer	Wine[34] / beer[35]
	Isobutanol	Wine, vinous		Wine / beer	Wine[36] / beer[37]
	Phenethyl alcohol*†	Floral, rose		Wine / beer	Wine[38]
Ketone	4-methylacetophenone	Floral, hawthorne	Grain	Wine	Wine[39]
	Diacetyl*†	Buttery	Fermentation	Wine / beer	Wine[40] / beer[41]
Lactone	*Trans*-whiskey lactone	Coconut, celery	Maturation	Wine	—
	Cis-whiskey lactone†	Coconut, oak		Wine	—
	Sotolon	Caramel, curry		Wine	—
	6-dodeceno-γ-lactone	Peach	Grain, fermentation	Wine	Wine[42]
	γ-decalactone	Coconut, peach		Wine / beer	Wine[43]
	γ-dodecalactone	Coconut, peach		Wine / beer	Wine[44]
	γ-nonalactone†	Coconut, peach		Wine / beer	Wine[45]
	δ-nonalactone	Peach		Wine	Wine[46]

(*continued*)

TABLE 10.1 CHARTING THE CHEMICAL ROADMAP OF TERROIR IN BOURBON (CONTINUED)

CLASSIFICATION	IMPORTANT FLAVOR COMPOUND IN BOURBON	AROMA	ORIGIN(S)	REPORTED IN WINE OR BEER	INFLUENCED BY TERROIR IN WINE OR BEER
Methoxypyrazine	2-isopropyl-3-methoxypyrazine	Earthy	Grain	Wine	Wine[47]
Organic acid	Phenylacetic acid	Honey, floral	Grain, fermentation	Wine / beer	Wine[48]
		Cooked apple	Grain	Wine	—
Terpene	α-damascone				
	β-damascenone*†	Cooked apple		Wine / beer	Wine[49]
	β-ionone	Violets		Wine / beer	Wine[50]
Thioether	Dimethyl sulfide	Cooked corn	Grain, fermentation	Wine / beer	Wine[51] / beer[52]
Volatile phenol	4-ethylguaiacol†	Phenolic, smoky, bacon	Grain, fermentation	Wine / beer	Wine[53]
	4-ethylphenol	Band-Aid, smoky		Wine / beer	Wine[54]
	Eugenol†	Clove	Grain, maturation	Wine / beer	Wine[55]
	Guaiacol†	Woody, smoky		Wine / beer	Wine[56]
	Vanillin†	Vanilla	Maturation	Wine / beer	—

*Denotes one of the eighteen flavor compounds responsible for the global vinous odor in wine. See V. Ferreira et al., "The Chemical Foundations of Wine Aroma—a Role Game Aiming at Wine Quality, Personality, and Varietal Expression," in *Proceedings of the Thirteenth Australian Wine Industry Technical Conference* (Adelaide: Australian Wine Industry Technical Conference, 2007).

†Denotes one of the twenty-five flavor compounds used in the aroma recombinate whose aroma was indiscernible from the bourbon sample. See L. Poisson and P. Schieberle, "Characterization of the Key Aroma Compounds in an American Bourbon Whisky by Quantitative Measurements, Aroma Recombination, and Omission Studies," *Journal of Agricultural and Food Chemistry* 56, no. 14 (2008): 5820–26.

Sources can be found in appendix 2: "Key to the Roadmap: Sources for Chapter 10."

These precursors are bound to sugars in the fruit but are released during fermentation and develop into these aromatic norisoprenoid terpenes. More sunlight in the vineyard appears to encourage the development of most norisoprenoid terpenes. Interestingly, this is the opposite of β-damascenone, which thrives under less sunlight.[5] Ultimately, I found over a dozen reports showing that the genetic and environmental factors that cumulatively make up terroir have an effect on the presence and concentration of norisoprenoid terpenes in wine (table 10.1). The same may hold true for their presence and concentration in whiskey.

While promising, the omission studies threw a curveball for the importance of β-damascenone. When an aroma recombinate *omitted* β-damascenone, it was indistinguishable from the original recombinate that contained all twenty-five flavor compounds. This is the opposite of what Ferreira found, which is that β-damascenone was one of two (the other was the banana-scented ester isoamyl acetate) from the global odor group that changed the aroma of wine when it was omitted. The wine mixture without β-damascenone had an overall less intense flavor. Schieberle noted that β-damascenone's impact could have been masked by the intensely fruity esters present in the aroma recombinate, but still, this was perplexing. It was the first finding of the paper that made me raise an eyebrow, questioning the sensory methods.

Of the eight lactones identified by Schieberle, at least three were derived either directly or indirectly (through fermentation precursors) from grain: γ-nonalactone, γ-decalactone, and γ-dodecalactone. All three of these lactones carry *coconut*, *peach*, and *creamy sweet* flavors. γ-nonalactone seemed to be the most promising, as it had the second highest FD factor, the twentieth highest OAV, and was included in the aroma recombinate (γ-decalactone and γ-dodecalactone were not). While γ-decalactone and γ-dodecalactone are produced by yeast from fatty acid precursors during fermentation, we believe γ-nonalactone is formed by the enzymatic oxidation of grain-derived linoleic acid during mashing.[6] Similar to β-damascenone, γ-nonalactone is an important flavor contributor in wine—they even call it a "hidden key wine odorant"[7] (an *odorant* is just another word for a volatile flavor compound), in that its concentration is low but its impact is high. And while its formation would not come from high-temperature-induced enzymatic activity in wine, its concentration is still affected by vineyard conditions[8] and grape variety.[9] However, just as with β-damascenone, when Schieberle and his team compared an aroma recombinate lacking γ-nonalactone to the original twenty-five-odorant recombinate, they couldn't tell the difference.

γ-nonalactone's close coconut-scented cousin *cis*-whiskey lactone also had a high FD factor and the ninth highest OAV. Whiskey lactone is largely recognized as one of the most important flavor contributors to oak-aged wine and whiskey. When Schieberle's team prepared an aroma recombinate without *cis*-whiskey lactone, it was easily and obviously distinguishable from the bourbon sample and the complete twenty-five-compound aroma recombinate.

The *cis*- moniker tells us that whiskey lactone has different isomers, which are compounds that share the same atoms—that is, they have the same *chemical formula*—but have a different arrangement of atoms in space. Imagine you have two sets of five pennies, two nickels, and one quarter on a desk. You arrange one set from top to bottom starting with the five pennies, then the two nickels, and then ending with the quarter at the base. You arrange the second set in the opposite direction, starting with the quarter at the top. These two sets are isomers of each other. They contain the same number and types of coins (atoms), but they are arranged differently in space. It may seem trivial, but the arrangement of atoms in space means that two isomers with the same chemical formula can have very different aroma characteristics or sensitivities.

Cis-whiskey lactone is present in higher concentrations and is more aroma-active than its isomer, *trans*-whiskey lactone. But they are both deemed important for flavor. That said, the whiskey lactones enter wine and whiskey during their maturation in oak barrels. So, while important to flavor, the two isomers have nothing to do with grains, grapes, or terroir.

Schieberle and his team detected eleven aldehydes with high FD factors, five of which were included in the aroma recombinate. Ten of these eleven aldehydes originate directly from grain precursors. Specifically, the high temperatures of the malting, mashing, and distillation process induce

Strecker degradations of amino acids and oxidation of lipids, which produce a variety of different aroma-active aldehydes. The eleventh aldehyde was acetaldehyde, which is produced by yeast as an intermediate in ethanol production. It is also formed from the oxidation of ethanol during maturation.

Of the aldehydes linked to lipid oxidation, 2,6-nonadienal (*cucumber*), 2-heptenal (*apple*), 2,4-decadienal (*fatty*), 2-nonenal (*stale*), and 2-nonenal's derivative nonanal (*soapy*) were the most interesting because of their high FD factors. These four compounds are all reported in wine, and their concentrations are affected by some aspect of terroir, from grape variety to vineyard conditions to vintage variations. For example, one report showed that 2,6-nonadienal had an FD factor in merlot wine double that of cabernet sauvignon and four times that of cabernet franc and cabernet gernischt.[10] I also discovered that these flavor compounds had been investigated in beer. Specifically, the presence of 2-heptenal and 2-nonenal in a crude beer mash depended on the variety of barley.[11]

The other group of aldehydes arose from Strecker degradations of amino acids: 2-methylbutanal (*chocolate*) from isoleucine, isobutyraldehyde (*grainy*) from valine, and isovaleraldehyde (*chocolate*) from leucine. All three contained high FD factors. The latter two also had high OAVs and were included in the aroma recombinate. Once again, I discovered that terroir could affect their presence and concentration in both wine and beer.

For example, a beer study found that among fourteen different beer mashes—each with their own specific variety of barley or wheat—2-methylbutanal was detected in only two of them.[12] Another found that while isobutyraldehyde and isovaleraldehyde were present in a selection of ten barley varieties, the concentrations were significantly higher in the French and Australian varieties as opposed to the Canadian and Chinese ones.[13] Maybe even more so than the norisoprenoid terpenes and lactones, certain aldehydes appeared to be prime whiskey terroir candidates. But then I read the omission studies by Schieberle's team, which showed that a twenty-three-odorant aroma recombinate without isobutyraldehyde and isovaleraldehyde was not distinguishable from the original recombinate. Again, I found this confounding: isovaleraldehyde had the second highest OAV, and isobutyraldehyde had the eighth. How could removing such seemingly foundational flavor compounds not change the aroma? I was perplexed—even discouraged—by this apparent contradiction. But I still felt that the roadmap to whiskey terroir was developing.

Next I considered a class of compounds called methoxypyrazines. While it wasn't included in the aroma recombinate, I was intrigued that Schieberle had identified 2-isopropyl-3-methoxypyrazine (IPMP) in the whiskey. It shared one of the lowest FD factors among the fifty and therefore is unlikely to be a major contributor to flavor in the assessed bourbon, but IPMP is an important contributor of *potato*, *earthy*, and *asparagus* aromas in certain wines. While some yeast species and strains—and even the lady beetle, a vineyard pest (*Harmonia axyridis*)[14]—can produce IPMP, its presence in wine is believed to derive primarily from the grape berry itself and its amino acid precursors. IPMP levels are reported to be highest in cabernet sauvignon grapes, and its levels depend on a combination of environmental factors including sunlight, temperature, humidity, and soil.[15] So while its concentration might have been too low to change Poisson and Schieberle's bourbon, the fact that it was identified suggests that it is produced in grain kernels and that potentially certain species and varieties—depending on the influence of growing environment—may possess levels high enough to contribute to a whiskey's flavor.

Schieberle identified five volatile phenols with high FD factors: eugenol (*clove*), 4-ethylguaiacol (*phenolic*, *smoky*, *bacon*), 4-ethylphenol (*Band-Aid*, *smoke*), guaiacol (*woody*, *smoky*), and vanillin (*vanilla*). All but 4-ethylphenol were incorporated into the aroma recombinate. In general, volatile phenols impart flavors that range from *smoky* and *medicinal* to *barnyard* and *sweaty saddle* to *vanilla* and *spice*. They are a diverse set of compounds, and the human nose is especially sensitive to their presence. Volatile phenols can come from fermentation byproducts; from malts that were dried using smoke (such as peat smoke or wood smoke); from thermal degradation of grain constituents during high-temperature malting, mashing, or distillation; and from oak maturation.

Like whiskey lactone, many of the volatile phenols found in bourbon are usually attributed to oak maturation produced from lignin degradation during toasting and charring. Schieberle's group found vanillin had the third highest FD factor and OAV. But vanillin is a good example of a volatile phenol whose origin is largely—if not solely—maturation. So, while vanillin matters loads for flavor, it is not a prime candidate for investigating the role of terroir in whiskey. But the four other volatile phenols could have origins in grain as well as in fermentation and maturation.

Similar to vanillin, guaiacol and eugenol do indeed derive from oak. In general, these two compounds are usually attributed to barrel maturation. But they can also derive from thermal degradation of lignin in grain. Researchers have found guaiacol and eugenol in wines that were not aged in barrels. Given that grape lignin breakdown can result in these flavor compounds in wine, it's reasonable to assume that grain lignin can provide them as well. And like the flavor compounds I had already investigated, multiple reports showed that the presence and concentration of guaiacol and eugenol in wine is influenced by grape variety, vineyard environment, and vintage.

4-ethylguaiacol and 4-ethylphenol are formed by the thermal degradation or microbial metabolism of 4-vinylguaiacol (*clove, spice*) and 4-vinylphenol (*medicinal, sweet*), respectively. While not identified in Schieberle's research, 4-vinylguaiacol and 4-vinylphenol are well-known flavor compounds in whiskey, beer, and wine. The characteristic clove, spice, and phenolic flavors of hefeweizens, witbiers, and saisons are largely caused by these two volatile phenols. They are created by the thermal degradation (during malting, mashing, and/or distillation) or microbial metabolism (during fermentation) of hydroxycinnamic acids, which are the building blocks of lignin and lignan. 4-vinylguaiacol is formed through the thermal degradation or microbial metabolism of ferulic acid, and 4-vinylphenol is formed through the degradation or metabolism of coumeric acid. Only certain yeast species and strains can produce the necessary enzymes to metabolize ferulic acid and coumeric acid into their vinyl phenol derivatives. The *Saccharomyces cerevisiae* yeast strains used to make hefeweizens, witbiers, and saisons typically have the genetic machinery to produce these necessary enzymes.

4-ethylguaiacol and 4-ethylphenol result from either the thermal degradation or microbial metabolism of their vinyl phenol precursors. If it's the former, it occurs during malting, mashing, or distillation. If it's the latter, it occurs during fermentation and is typically attributed to the wild yeast *Brettanomyces*, which can produce the enzymes necessary to metabolize 4-vinylguaiacol (*clove, spice*) and 4-vinylphenol (*medicinal, sweet*) into 4-ethylguaiacol and 4-ethylphenol, respectively.

Once again, multiple reports—many of the same that investigated guaiacol and eugenol—also showed that 4-vinylguaiacol, 4-ethylguaiacol, 4-vinylphenol, and 4-ethylphenol can be influenced by terroir in wine. Further, a report published about a year before Schieberle's showed that the hydroxycinnamic acid precursors for 4-vinylguaiacol and 4-vinylphenol varied widely based on the variety of barley, where the barley grew, and the agronomic techniques employed.[16]

Finally, I turned my attention to what is arguably the most important class of flavor compounds in wine, beer, and whiskey—esters. Of the fifty flavor compounds detected with high FD factors, fourteen were esters. Nine of those fourteen had OAVs above 1 and were included in the twenty-five-compound aroma recombinate. When the ethyl esters and isoamyl acetate were omitted, the derivative aroma recombinate was easily distinguishable from the original.

Esters are not present in high enough concentrations in grapes or grains to change the flavor of wine or whiskey. Instead, those esters so crucial to flavor—largely contributing fruity and floral aromas—are produced by yeast during fermentation. This largely occurs in the cytoplasm of a yeast cell from enzymatic condensation reactions—which we call *esterification*—of organic acids and alcohols (both ethanol and fusel alcohols). A condensation reaction is one in which two compounds merge together and release a water molecule in the process. The precursor organic acids and alcohols of esters are themselves produced by upstream metabolic pathways in yeast that start with sugars and amino acids. Different profiles of sugars (glucose versus maltotriose versus fructose, for example) and amino acids in the must, wort, or mash can produce different organic acids and alcohols, which then produce different profiles of esters.[17] Once produced, esters are expelled from the yeast cell into the fermenting wine or beer. Reports also show that ester production in wine and whiskey is not just linked to yeast metabolism per se. For example, it has been shown that bacteria such as *Oenococcus oeni* (which has been identified in fuel ethanol fermentations)[18] and *Lactobacillus plantarum* (which has been identified in whiskey fermentations)[19] can produce esters.[20]

Besides synthesizing esters, bacteria produce prodigious amounts of organic acids—namely, lactic and acetic acids—during whiskey mash fermentation. These acids can serve as precursors for esterification reactions that occur via chemical—not biological—means in beer. The persistence of bacteria in whiskey mash fermentations is not because the distiller didn't clean the tank (or at least not usually). Unlike beer production, where the wort is boiled to kill anything the brewer doesn't want, the cooking of rye, wheat, barley, and malt usually occurs at temperatures low enough (145°F/62°C to 155°F/68°C) for some of the indigenous bacteria on the grain to survive. (Corn is cooked at higher temperatures, usually around 190°F/87°C and above, as it requires higher heat to gelatinize the starch. At these temperatures, most of the indigenous bacteria are killed.)

Many distilleries, including my own, encourage a "late lactic fermentation" once the yeast have moved past their active growth phase. This allows bacteria (largely *Lactobacillus*) to feed on residual nutrients and autolyzing (essentially, self-destructing) yeast cells.[21] The makeup of *Lactobacillus* species and strains is specific and unique to individual distilleries, which appears to contribute to "house flavors" in whiskey.[22]

The majority of the esters detected by Schieberle's team with high FD factors and OAVs were ethyl esters, which derive from the condensation of ethanol and fatty acids. In descending order, ethyl-2-methylbutyrate (*apple*), ethyl hexanoate (*apple*), ethyl butyrate (*pineapple*), and ethyl octanoate (*fruity, floral*) scored the highest OAVs of all the measured esters. There were also some important acetate esters, which come from the condensation of acetic acid with either ethanol (which solely results in ethyl acetate) or fusel alcohols (which results in a number of different ester possibilities). The two acetate esters identified with OAVs above 1 and thus included in the aroma recombinate were phenylethyl acetate (*honey*) and isoamyl acetate (*banana*).

A high concentration of these two acetate esters isn't surprising: phenethyl alcohol (*rose*) and isoamyl alcohol (*banana*)—the precursor fusel alcohols for phenylethyl acetate and isoamyl acetate, respectively—were also identified as having high OAVs and were included in the aroma recombinate.

All fourteen of the esters and three of the fusel alcohols Schieberle found are reported to be present in wine and influenced by grape variety, vineyard environment, and agronomic techniques. This meant terroir's influence in whiskey would likely lie in how different grain varieties and growing environments changed the composition and concentration of the basal fermentation precursors—amino acids and sugars. But we also know that ester production depends greatly on the species and strains of microbes and the fermentation conditions, such as temperature and length.

* * *

As I dug deeper and explored all fifty of the most odor-active flavor compounds in the bourbon, I discovered that every single one was also present in wine and/or beer. And most importantly, I found ample research reports (all of which are cited in appendix 2, which contains the notes for table 10.1, if you would like to read any of the original research for yourself) showing that those compounds derived from grain in the bourbon—either directly through thermal degradations and oxidations or indirectly as fermentation precursors—were affected by some aspect of terroir in wine, beer, or both.

I thought back to Reid Mitenbuler's book *Bourbon Empire: The Past and Future of America's Whiskey*. One key whiskey character that book covers is Dr. James Crow, one of the first distillers to bring modern science to bourbon production in the first half of the nineteenth century. Crow had "an obsession with improving his whiskey," according to Mitenbuler. In true scientist fashion, Crow recorded more variables than arguably anyone before him. He used instruments to measure sugar and alcohol levels, investigated the importance of using water that was low in iron, monitored and controlled fermentation temperatures, and strove for balance and control as "he navigated his whiskey through a tangled jungle of factors." In the complexity of its production, Crow realized whiskey "wasn't just a straight line—beer in, alcohol out—it was a quilt stitched together in interconnected webs. You pull on one tiny part, and the whole thing shifts."[23] Terroir, I was beginning to discover, pulled at many threads.

Schieberle's research had given me the means to chart a roadmap for terroir in bourbon. And terroir appeared to be a phenomenon that could change the presence and concentration of a wide range of flavor compounds that we find in the finished whiskey. But this raised an important point. Bourbon is just one style, and it's made from very different grains than many of the other popular whiskey styles. What would the roadmap of terroir look like for rye whiskey or malt whiskey?

11

OVERLAYING THE MAPS

WINE AND BOURBON may look and taste different, but there is an impressive amount of overlap between the chemical compounds that make up their flavor. Through aroma extract dilution analysis (AEDA) and the subsequent calculations of flavor dilution (FD) factors and odor activity values (OAVs), scientists identified fifty flavor compounds that are critically important in bourbon. And all fifty of those have been identified as flavor compounds in wine, beer, or both. Thirteen of them were also reportedly contributors to the global vinous odor of wine.

Seeing the extent to which these compounds overlapped allowed me to begin producing my chemical roadmap to the terroir of whiskey. The research on wine is extensive, and the vast majority of the fifty potentially important flavor compounds in bourbon—at least those that derive from the grain (either directly from thermal degradations and oxidations or indirectly as a fermentation precursors)—are meaningfully influenced in wine by grape variety, vineyard environment, or agronomic technique.

However, so far, bourbon was the only style of whiskey I'd used to develop the roadmap, and only one brand, at that. Would these fifty flavor compounds be present in other styles? And, if so, would they be equally influential? I had to discover whether any other whiskey styles had been investigated using a sensomics approach. If they hadn't, my chemical roadmap to the terroir of whiskey would really just be a roadmap to the terroir of bourbon. Given that bourbon is produced only in the United States, that would leave most of the whiskey-making world out of the loop, and that wouldn't suffice.

In 2010, a scientist named Jacob Lahne submitted his master's thesis, "Aroma Characterization of American Rye Whiskey by Chemical and Sensory Assays," at the University of Illinois at Urbana-Champaign's food sciences program. He employed many of the same techniques pioneered by Schieberle to identify the most important flavor compounds in two different Kentucky rye whiskeys.[1] Unlike the bourbon sensomics papers, Lahne revealed which brands he had used: Rittenhouse and Wild Turkey. Both brands are well established, and they both use a grain bill where rye is at the legal minimum of 51 percent. Like the bourbon assessed by Schieberle, these whiskies would have used modern varieties of rye sourced from the commodity system. Indeed, most Kentucky distilleries are still sourcing rye from Europe, the northern Midwest, or Canada. Because the grains travel long distances, they're blended over and over, in elevators, cargo ships, railcars, and truck hoppers. While there are ongoing projects to grow rye in Kentucky, it will be many years (if ever) before the majority of Kentucky rye whiskey is made from Kentucky rye.

Using the same sensomics techniques pioneered by Schieberle and his colleagues at the Technical University of Munich group, Lane identified thirty-one flavor compounds that were necessary to create the flavor of both rye whiskeys, and he incorporated them all into his own aroma recombinate. I've merged the chemical roadmap of terroir in bourbon (table 10.1) with the chemical roadmap of terroir in rye whiskey (table 11.1). As in chapter 10, table 11.1 will help you follow along with the results relayed in this chapter.

Only eight of the thirty-one compounds were not also identified as important contributors to flavor in the bourbon assessed in the Schieberle papers. Of those eight, four were volatile phenols: 4-vinylguaiacol (*spice*,

TABLE 11.1 CHARTING THE CHEMICAL ROAD MAP OF TERROIR IN BOURBON AND RYE WHISKEY

CLASSIFICATION	IMPORTANT FLAVOR COMPOUND IN BOURBON[1]	IMPORTANT FLAVOR COMPOUND IN RYE WHISKEY[2]	AROMA	ORIGIN(S)	INFLUENCED BY TERROIR IN WINE OR BEER
Acetal	1,1-diethoxyethane	—	Sweet, green, ethereal	Fermentation, maturation	Wine[3]
Aldehyde	2,4-nonadienal	—	Fatty, melon	Grain	—
	2,6-nonadienal	—	Cucumber		Wine[4] / beer[5]
	2-decenal	—	Orange, floral		—
	2-heptenal	—	Apple, fatty		Wine[6] / beer[7]
	2-nonenal	—	Stale bread, cardboard		Wine[8] / beer[9]
	2-methylbutanal	—	Chocolate		Wine[10] / beer[11]
	2,4-decadienal	—	Meaty, fatty		Wine[12]
	Isobutyraldehyde	—	Grainy		Wine[13] / beer[14]
	Isovaleraldehyde	—	Chocolate		Wine[15] / beer[16]
	Nonanal	—	Soapy, fatty		Wine[17]
	Acetaldehyde*	Acetaldehyde*	Green apple	Fermentation, maturation	Wine[18] / beer[19]
Ester	2-phenethyl propionate	—	Floral, rose, sweet	Fermentation	—
	Ethyl 2-methylbutyrate*	—	Apple		Wine[20]
	Ethyl 2-phenylacetate	—	Cocoa, honey, floral		Wine[21]
	Ethyl butyrate*	Ethyl butyrate*	Pineapple		Wine[22]

	Ethyl cinnamate	Ethyl cinnamate	Cinnamon		Wine[23]
	Ethyl hexanoate*	Ethyl hexanoate*	Fruity, apple		Wine[24]
	Ethyl isobutyrate*	Ethyl isobutyrate*	Citrus, strawberry		Wine[25]
	Ethyl isovalerate*	Ethyl isovalerate*	Apple, pineapple		Wine[26]
	Ethyl octanoate*	—	Fruity, floral, banana, pineapple		Wine[27]
	Ethyl pentanoate	—	Apple, pineapple		Wine[28]
	Ethyl propanoate	Ethyl propanoate	Grape		Wine[29]
	Isoamyl acetate*	Isoamyl acetate*	Banana		Wine[30]
	Phenylethyl acetate	Phenylethyl acetate	Honey		Wine[31] / beer[32]
	Ethyl acetate*	—	Fruity, ethereal	Fermentation, maturation	Wine[33] / beer[34]
	—	Ethyl vanillate	Vanilla, sweet		Wine[35]
Fusel alcohol	Isoamyl alcohol*	Isoamyl alcohol*	Banana	Fermentation	Wine[36] / beer[37]
	Isobutanol	Isobutanol	Wine, vinous		Wine[38] / beer[39]
	Phenethyl alcohol*	Phenethyl alcohol*	Floral, rose		Wine[40]
Ketone	4-methylacetophenone	—	Floral, hawthorne	Grain	Wine[41]
	Diacetyl*	—	Buttery	Fermentation	Wine[42] / beer[43]
Lactone	*trans*-whiskey lactone	*trans*-whiskey lactone	Coconut, celery	Maturation	—
	cis-whiskey lactone	*cis*-whiskey lactone	Coconut, oak		—
	Sotolon	—	Caramel, curry		—

(*continued*)

TABLE 11.1 CHARTING THE CHEMICAL ROAD MAP OF TERROIR IN BOURBON AND RYE WHISKEY (CONTINUED)

CLASSIFICATION	IMPORTANT FLAVOR COMPOUND IN BOURBON	IMPORTANT FLAVOR COMPOUND IN RYE WHISKEY	AROMA	ORIGIN(S)	INFLUENCED BY TERROIR IN WINE OR BEER
	6-dodeceno-γ-lactone	—	Peach	Grain, fermentation	Wine[44]
	γ-decalactone	—	Coconut, peach		Wine[45]
	γ-dodecalactone	—	Coconut, peach		Wine[46]
	γ-nonalactone	γ-nonalactone	Coconut, peach		Wine[47]
	δ-nonalactone	—	Peach		Wine[48]
Methoxypyrazine	2-isopropyl-3-methoxypyrazine	—	Earthy	Grain	Wine[49]
Organic acid	—	Acetic acid	Vinegar	Fermentation, maturation	—
	—	Butyric acid*	Rancid	Fermentation	Wine[50]
	—	Isovaleric acid*	Cheesy		Wine[51]
	Phenylacetic acid	Phenylacetic acid	Honey, floral	Grain, fermentation	Wine[52]
Terpene	α-damascone	—	Cooked apple	Grain	—
	β-damascenone*	β-damascenone*	Cooked apple		Wine[53]
	β-ionone	β-ionone	Violets		Wine[54]

Thioether	Dimethyl sulfide	—	Cooked corn	Grain, fermentation	Wine[55] / beer[56]
Volatile phenol	—	4-vinylguaiacol	Spice, clove	Grain, fermentation	Wine[57]
	4-ethylguaiacol	4-ethylguaiacol	Phenolic, smoky, bacon		Wine[58]
	4-ethylphenol	4-ethylphenol	Band-Aid, smoky		Wine[59]
	Eugenol	Eugenol	Clove	Grain, maturation	Wine[60]
	Guaiacol	Guaiacol	Woody, smoky		Wine[61]
	Vanillin	Vanillin	Vanilla	Maturation	—
	—	Syringaldehyde	Sweet, green		—
	—	Syringol	Sweet, smoky		—
	—	*p*-cresol	Band-Aid	Grain	—

*Denotes one of the eighteen flavor compounds responsible for the global vinous odor in wine. See V. Ferreira et al., "The Chemical Foundations of Wine Aroma—a Role Game Aiming at Wine Quality, Personality, and Varietal Expression," in *Proceedings of the Thirteenth Australian Wine Industry Technical Conference* (Adelaide: Australian Wine Industry Technical Conference, 2007).

Sources can be found in appendix 3: "Key to the Roadmap: Sources for Chapter 11."

clove), cresol (*Band-Aid*), syringaldehyde (*sweet, green*), and syringol (*sweet, smoky*); three were organic acids: acetic acid (*vinegar*), butyric acid (*rancid*), and isovaleric acid (*cheesy*); and one was an ester: ethyl vanillate (*vanilla*).

So I now had fifty-eight whiskey flavor compounds, twenty-three of which were important in both bourbon and rye.

Looking at the list, at first I was confused. For example, the source of syringaldehyde and syringol is usually attributed to the oak barrel, derived from lignin and the thermal degradation of its syringyl building blocks when the oak is charred. Bourbon and rye are both aged in the same type of barrel—new, charred oak. So why weren't syringaldehyde and syringol not also identified in Schieberle's research? The presence of cresol is also a bit of a headscratcher, as this compound is usually attributed to the use of malt that has been kilned and infused with peat[2] or to the use of heavily roasted malts in beer.[3]

But I kept investigating, and I found the results to be both encouraging and clarifying. Consider 4-vinylguaiacol. I have already discussed how this flavor compound is actually a precursor to 4-ethylguaicol (*phenolic, smoky, bacon*), which was indeed identified—along with 4-ethylphenol (*Band-Aid, smoky*)—in the bourbon. Lahne also found 4-ethylguaicol, 4-ethylphenol, eugenol (*clove*), and guaiacol (*woody, smoky*)—all of which were important flavor compounds in Schieberle's bourbon—in the rye whiskeys. So although not a perfect overlap, very similar profiles of volatile phenols were shared between the two whiskey styles.

But building on these similarities and differences, can we develop hypotheses to explain them? I think we can. For example, I think it is a sound hypothesis to say that rye whiskey (made primarily from rye) could contain different or increased concentrations of certain lignin- and lignan-derived volatile phenols and fewer lipid-derived aldehydes than bourbon (made primarily from corn).

Schieberle identified ten aldehydes from grain in bourbon that were important to its flavor. Seven of these ten are the result of lipid oxidations that occur during malting, mashing, or distillation. Table 11.2 shows that corn has more than double the amount of fat (a type of lipid) as rye. So the presence of more fat in corn might explain why there are more aldehydes

TABLE 11.2 VARYING LEVELS OF CHEMICAL COMPONENTS AMONG THE FOUR MAJOR WHISKEY GRAINS

COMPONENT	CORN	RYE	WHEAT	BARLEY
Starch (percent w/w)	65.0	60.3	59.4	62.7
Protein (percent w/w)	8.8	9.4	11.3	11.1
Fat (percent w/w)	3.8	1.7	1.8	2.1
Fiber (percent w/w)	9.8	13.1	13.2	9.7
Lignin (percent w/w)	1.4	2.1	2.0	1.9
Lignans (μg/100 g)	23	770	76	205

Source: P. Koehler and H. Wieser, "Chemistry of Cereal Grains," in *Handbook on Sourdough Biotechnology*, ed. M. Gobbetti and M. Gänzle (New York: Springer, 2013), 11–45.

in bourbon. And of the ten grain-derived aldehydes found in bourbon by Schieberle, not one was detected in the rye whiskeys by Lahne.

For those volatile phenols that Schieberle and Lahne both found, the concentrations were higher in the rye whiskey (table 11.3). Why? Dr. Don Livermore, the master blender at Hiram Walker and a close colleague of mine, clarified this for me while I was researching this topic. He pointed out that rye has significantly higher levels of lignin and the closely related constituent lignans than corn (table 11.2). The thermal degradation of the lignin and lignan during mashing and distillation can produce volatile phenols. This is partly why we taste rye whiskey as spicier and more phenolic than bourbon, Dr. Don explained. So another sound hypothesis to describe variations in the presence and concentration of certain volatile phenols in rye whiskey versus bourbon whiskey can be tied to the concentration of lignin and lignan in the grains.

The presence of the acetic, butyric, and isovaleric organic acids in the rye whiskeys but not the bourbon might be because of rye's lower mashing temperatures. In a bourbon mash, the corn is cooked at temperatures between 190°F (88°C) and 212°F (100°C) to effectively gelatinize its crystalline starch. In a rye whiskey mash, the rye is usually cooked at a much lower temperature—148°F (64°C) to 155°F (68°C)—as its starch is not as tough to gelatinize. The hot temperatures used to cook corn in a bourbon mash will

TABLE 11.3 CONCENTRATION DIFFERENCES FOR THREE VOLATILE PHENOLS BETWEEN BOURBON AND RYE WHISKEY

CLASSIFICATION	FLAVOR COMPOUND	CONCENTRATION IN BOURBON (PARTS PER BILLION)[a]	CONCENTRATION IN WILD TURKEY RYE (PARTS PER BILLION)[a]	CONCENTRATION IN RITTENHOUSE RYE (PARTS PER BILLION)[a]
Volatile phenol	4-ethylguaicol	59	2,180	187
	Eugenol	240	583	993
	Guaiacol	56	3,760	3,150

[a]J. Lahne, "Aroma Characterization of American Rye Whiskey by Chemical and Sensory Assays," master's thesis, University of Illinois at Urbana-Champaign, 2010, http://hdl.handle.net/2142/16713.

kill most of the vegetative and sporulating bacteria. However, many bacteria—including both non-spore-forming (such as *Lactobacillus*) and spore-forming (such as *Clostridium*) species—survive the temperatures of a rye mash.[4] So rye whiskey fermentations could theoretically contain higher concentrations of bacteria, which produce more organic acids.

Ultimately, there were eight flavor compounds present in the rye whiskeys that were not present in the bourbon. And I had some hypotheses supported by the scientific literature and my experience with whiskey making to explain why. But there were twenty-three important flavor compounds shared by both. Three of them were likely either mostly or wholly from the oak barrel: the two whiskey lactones and vanillin. Of the remaining twenty flavor compounds derived from grain—either directly from thermal degradations and oxidations or indirectly as fermentation precursors—fourteen of them had OAVs above 1 in Schieberle's bourbon. These were the norisoprenoid terpene β-damascenone (*cooked apple*); the esters ethyl cinnamate (*cinnamon*), ethyl isovalerate (*apple, pineapple*), ethyl hexanoate (*apple*), ethyl isobutyrate (*citrus, strawberry*), ethyl butyrate (*pineapple*), phenylethyl acetate (*honey*), and isoamyl acetate (*banana*); the lactone γ-nonalactone (*coconut, peach*); the fusel alcohols phenethyl alcohol (*rose*) and isoamyl alcohol (*banana*); and the volatile phenols 4-ethylguaiacol (*phenolic, smoky, bacon*), eugenol (*clove*), and guaiacol (*woody, smoky*).

The last group of six grain-derived flavor compounds therefore belonged to a group that possessed high FD factors in Schieberle's bourbon sample but whose concentration was measured below their odor threshold: the aldehyde acetaldehyde (*green apple*), the ester ethyl propanoate (*grape*), the fusel alcohol isobutanol (*wine, vinous*), the norisoprenoid terpene β-ionone (*violets*), the organic acid phenylacetic acid (*honey*), and the volatile phenol 4-ethylphenol (*Band-Aid, smoky*). According to Lane's research, all six of these compounds were present above their odor thresholds and necessary for the creation of an effective aroma recombinate.

This may appear to complicate matters. But perhaps instead it raises an important point about what distinguishes rye whiskey from bourbon—and even from all other whiskeys, for that matter. Instead of the two styles possessing vastly different flavor compounds, maybe their differences are driven more by concentration differences among the same or very similar compounds. They may all traverse the same general roadmap, but the specific roads they cross, and how often they cross them, is specific to the styles.

This is the case in wine. Terroir pulls at many threads in the flavor tapestry. Sometimes this pull results in unique flavor compounds specific to one varietal or growing region. But the pull of terroir can also lead to concentration differences in the flavor compounds that are present in many wines. If terroir pulls at wine's flavor tapestry in this way, then why wouldn't it do the same in whiskey?

After all, the origin of these shared flavor compounds hadn't changed—β-ionone, for example, still derived from grain carotenoids in the rye whiskeys just as it did in the bourbon sample. But some aspect of the diverging production processes or grain bills—which could be the use of different species and varieties from specific growing environments—caused the higher concentration of β-ionone in the rye.

I found this overarching hypothesis more promising than confounding. And perhaps, after all, we wouldn't need a different roadmap for every whiskey style. One map might be enough, with specific roads for those flavor compounds that were unique to a style, a region, or even a distillery.

That last differentiator—"distillery"—is a critical point to consider. The different grains used to make different styles of whiskeys are indeed responsible for much of the flavor variation. But what about brands of whiskeys of the same styles, some of which are produced from identical grain bills? Following the same reasoning, concentration differences among the same or very similar flavor compounds would most likely be responsible for the flavor nuances that exist in each brand among whiskeys of the same style, made from very similar ingredients, with very similar processes. For example, while 4-ethylguaiacol almost certainly contributed to aroma in both of Lahne's rye whiskeys, its OAV in Wild Turkey was 316, with a concentration of 2,180 parts per trillion (micrograms per liter), whereas its OAV was only 27, with a concentration of 187 parts per trillion, in Rittenhouse. These divergences—and the flavor nuances they would create—could be driven by the distiller pulling just a few threads of the thousands that compose the tapestry. This would seem to be why Wild Turkey rye and Rittenhouse rye—both made with identical grain bills and processed through very similar methods—possessed different OAVs (sometimes greatly different) for the same flavor compounds. And ultimately why they taste different.

* * *

In 2019, the Poznan University food scientist Henryk Jeleń and his colleagues used sensomic techniques to investigate a heavily peated scotch single malt whisky.[5] Like Schieberle's bourbon, Jeleń did not reveal which whisky his team analyzed. But regardless of the brand, just like the bourbon and rye whiskeys, the assessed scotch would have used modern barley varieties that were bred for yield and commodity-malt-quality parameters.

When I first discovered this study and found that it was the only sensonomic study on malt whiskey, I had mixed feelings. On the one hand, it was promising that with this paper I would now have sensomics data on three whiskeys made from the three dominant grains used in whiskey production: corn (in bourbon), rye (in rye whiskey), and barley (in malt whiskey). That said, *peated* malt whiskey (which dominates the whisky styles made on Scotland's island of Islay) is its own animal, drastically different in flavor from the fruitier, floral, and less intense whiskies made from unpeated malt that are typically found in the other four scotch regions: Speyside, Highlands, Lowlands, and Campbeltown. The medicinal, smoky, rubbery, and meaty flavors of peated scotch whiskies are intense and polarizing. The malt used in peated malt whiskey is kilned with peat—partially decayed hard-packed vegetable turf found in the bogs of Scotland—which leads to the infusion of phenolic compounds into the malt carried by the peat smoke. I feared the roadmap for peated whiskey would look very different than that of malt whiskies made from unpeated malt, not to mention bourbon and rye whiskey.

My concerns, though, were short-lived. Of the twenty-one compounds identified as key flavor compounds in the peated scotch malt whisky, fifteen were also key compounds in the bourbon and rye whiskeys. I merged the roadmaps once again, and the results are presented in table 11.4.

More and more, it looked like the flavor nuances among styles and brands were largely coming from the variations in concentration of the same (or very similar) compounds combined with a lesser number of unique compounds (what Ferreira calls *impact compounds* in wine; see chapter 3) limited to particular styles. Jeleń's paper did not report concentrations, but it did report FD factors. In general, Jeleń's peated malt whisky had higher FD factors for volatile phenols compared to bourbon and rye whiskeys. For example, 4-ethylphenol possessed an FD factor of 256 in the peated malt whisky but only 32 in the bourbon. This would mean that when the bourbon was diluted 1:32 with a neutral solvent, the aroma of 4-ethylphenol could just barely be identified by GC-O. But even if the peated malt whisky was diluted 1:256, it would still carry the phenolic and medicinal smell of 4-ethylphenol.

The volatile phenols guaiacol and cresol had the highest FD factors of all twenty-one flavor compounds measured in Jeleń's peated malt whisky. Guaiacol possessed an FD factor of 2,048 in the peated malt whisky (the highest among all twenty-one flavor compounds identified) and an FD factor of only 64 in the bourbon. Cresol was one of the thirty-one important flavor compounds in rye whiskey, but it had the lowest FD factor of the lot. The prominence of volatile phenols in the peated malt whisky as opposed

TABLE 11.4 CHARTING THE CHEMICAL ROAD MAP OF TERROIR IN BOURBON, RYE WHISKEY, AND MALT WHISKY

CLASSIFICATION	IMPORTANT FLAVOR COMPOUND IN BOURBON[62]	IMPORTANT FLAVOR COMPOUND IN RYE WHISKEY[63]	IMPORTANT FLAVOR COMPOUND IN MALT WHISKY[64]	AROMA	ORIGIN(S)	INFLUENCED BY TERROIR IN WINE OR BEER
Acetal	1,1-diethoxyethane†	—	1,1-diethoxyethane†	Sweet, green, ethereal	Fermentation, maturation	Wine[65]
Aldehyde	2,4-nonadienal	—	—	Fatty, melon	Grain	—
	2,6-nonadienal	—	—	Cucumber		Wine[66] / beer[67]
	2-decenal	—	—	Orange, floral		—
	2-heptenal†	—	2-heptenal†	Apple, fatty		Wine[68] / beer[69]
	2-nonenal†	—	—	Stale bread, cardboard		Wine[70] / beer[71]
	2-methylbutanal	—	—	Chocolate		Wine[72] / beer[73]
	2,4-decadienal†	—	—	Meaty, fatty		Wine[74]
	Isobutyraldehyde	—	—	Grainy		Wine[75] / beer[76]
	Isovaleraldehyde	—	—	Chocolate		Wine[77] / beer[78]
	Nonanal†	—	—	Soapy, fatty		Wine[79]
	Acetaldehyde*	Acetaldehyde*	—	Green apple	Fermentation, maturation	Wine[80] / beer[81]

Ester	—	—	Ethyl laurate†	Floral, waxy	Fermentation	Wine[82]
	—	—	Ethyl undecanoate†	Soapy, waxy		—
	2-phenethyl propionate	—	—	Floral, rose, sweet		—
	Ethyl 2-methylbutyrate*	—	—	Apple		Wine[83]
	Ethyl 2-phenylacetate	—	—	Cocoa, honey, floral		Wine[84]
	Ethyl butyrate*	Ethyl butyrate*	—	Pineapple		Wine[85]
	Ethyl cinnamate	Ethyl cinnamate	—	Cinnamon		Wine[86]
	Ethyl hexanoate*†	Ethyl hexanoate*†	—	Fruity, apple		Wine[87]
	Ethyl isobutyrate*	Ethyl isobutyrate*	—	Citrus, strawberry		Wine[88]
	Ethyl isovalerate*	Ethyl isovalerate*	Ethyl isovalerate*	Apple, pineapple		Wine[89]
	Ethyl octanoate*†	—	—	Fruity, floral, banana, pineapple		Wine[90]
	Ethyl pentanoate	—	—	Apple, pineapple		Wine[91]
	Ethyl propanoate	Ethyl propanoate	—	Grape		Wine[92]
	Isoamyl acetate*†	Isoamyl acetate*†	Isoamyl acetate*†	Banana		Wine[93]
	Phenylethyl acetate	Phenylethyl acetate	—	Honey		Wine[94] / beer[95]
	Ethyl acetate*†	—	—	Fruity, ethereal	Fermentation,	Wine[96] / beer[97]
	—	Ethyl vanillate	—	Vanilla, sweet	maturation	Wine[98]
Fusel alcohol	—	—	Isopropyl alcohol	Alcohol, solvent	Fermentation	Wine[99]
	Isoamyl alcohol*	Isoamyl alcohol*	Isoamyl alcohol*	Banana		Wine[100] / beer[101]
	Isobutanol	Isobutanol	Isobutanol	Wine, vinous		Wine[102] / beer[103]
	Phenethyl alcohol*†	Phenethyl alcohol*†	Phenethyl alcohol*†	Floral, rose		Wine[104]
Ketone	4-methylacetophenone	—	—	Floral, hawthorne	Grain	Wine[105]
	Diacetyl*	—	—	Buttery	Fermentation	Wine[106] / beer[107]

(*continued*)

TABLE 11.4 CHARTING THE CHEMICAL ROAD MAP OF TERROIR IN BOURBON, RYE WHISKEY, AND MALT WHISKY (CONTINUED)

CLASSIFICATION	IMPORTANT FLAVOR COMPOUND IN BOURBON[62]	IMPORTANT FLAVOR COMPOUND IN RYE WHISKEY[63]	IMPORTANT FLAVOR COMPOUND IN MALT WHISKY[64]	AROMA	ORIGIN(S)	INFLUENCED BY TERROIR IN WINE OR BEER
Lactone	*trans*-whiskey lactone	*trans*-whiskey lactone	*trans*-whiskey lactone	Coconut, celery	Maturation	—
	cis-whiskey lactone	*cis*-whiskey lactone	*cis*-whiskey lactone	Coconut, oak		—
	Sotolon	—	—	Caramel, curry		—
	6-dodeceno-γ-lactone	—	—	Peach	Grain, fermentation	Wine[108]
	γ-decalactone	—	γ-decalactone	Coconut, peach		Wine[109]
	γ-dodecalactone	—	—	Coconut, peach		Wine[110]
	γ-nonalactone	γ-nonalactone	—	Coconut, peach		Wine[111]
	δ-nonalactone	—	—	Peach		Wine[112]
Methoxypyrazine	2-isopropyl-3-methoxypyrazine	—	—	Earthy	Grain	Wine[113]
Organic acid	—	Acetic acid	—	Vinegar	Fermentation, maturation	—
	—	Butyric acid*	—	Rancid	Fermentation	Wine[114]
	—	Isovaleric acid*	—	Cheesy		Wine[115]
	Phenylacetic acid	Phenylacetic acid	—	Honey, floral	Grain, fermentation	Wine[116]

Terpene	α-damascone	—	—	Cooked apple	Grain	—
	β-damascenone*	β-damascenone*	—	Cooked apple		Wine[117]
	β-ionone	β-ionone	—	Violets		Wine[118]
Thioether	Dimethyl sulfide	—	—	Cooked corn	Grain, fermentation	Wine[119] / beer[120]
Volatile phenol	—	—	Benzenol	Phenolic	Grain	—
	—	—	4-ethyl-2-methyl phenol	Phenolic		—
	—	*p*-cresol	*p*-cresol	Band-Aid		—
	—	—	4-propylguaiacol	Clove		
	—	4-vinylguaiacol	4-vinylguaiacol	Spice, clove	Grain,	Wine[121]
	4-ethylguaiacol	4-ethylguaiacol	4-ethylguaiacol	Phenolic, smoky, bacon	fermentation	Wine[122]
	4-ethylphenol	4-ethylphenol	4-ethylphenol	Band-Aid, smoky		Wine[123]
	Eugenol	Eugenol	—	Clove	Grain,	Wine[124]
	Guaiacol	Guaiacol	Guaiacol	Woody, smoky	maturation	Wine[125]
	Vanillin	Vanillin	—	Vanilla	Maturation	—
	—	Syringaldehyde	—	Sweet, green		—
	—	Syringol	—	Sweet, smoky		—

*Denotes one of the eighteen flavor compounds responsible for the global vinous odor in wine. See V. Ferreira et al., "The Chemical Foundations of Wine Aroma—a Role Game Aiming at Wine Quality, Personality, and Varietal Expression," in *Proceedings of the Thirteenth Australian Wine Industry Technical Conference* (Adelaide: Australian Wine Industry Technical Conference, 2007).

†Denotes one of the thirty-six flavor compounds previously identified to be affected by terroir in whiskey. See R. J. Arnold et al., "Assessing the Impact of Corn Variety and Texas Terroir on Flavor and Alcohol Yield in New-Make Bourbon Whiskey," *PloS One* 14, no. 8 (2019).

Sources can be found in appendix 3: "Key to the Roadmap: Sources for Chapter 11."

to bourbon and rye whiskey is expected—that's exactly what the peat imparts! And importantly for our concept of terroir, studies show that the origin of the peat used in the malting process plays a role in the flavors it will impart to the malt and the eventual whiskey.[6]

Considering maturation, the whiskey lactones are some of the most important oak-derived flavor compounds. American whiskeys aged in new, charred American oak barrels are reported to possess higher concentrations than whiskeys that are aged in used barrels, like scotch malt whisky. While both the bourbon and the peated malt whisky contained whiskey lactones, as expected, the FD factors were quite different. In the bourbon, it was 1,024. In the peated malt whisky, it was only 64.

Only six flavor compounds identified in the peated malt whisky were not equally identified in the bourbon and rye whiskeys: the esters ethyl laurate (*floral*, *waxy*) and ethyl undecanoate (*soapy*, *waxy*); the fusel alcohol isopropyl alcohol (*alcohol*, *solventy*); and the volatile phenols benzenol (*phenolic*), 4-ethyl-2-methylphenol (*phenolic*), and 4-propylguaiacol (*clove*). The presence of unique esters is not surprising or discouraging. We touched on this previously. Many factors—yeast strain, fermentation conditions, the bacterial microbiome—influence the production of ethyl esters derived from long-chain fatty acids such as dodecanoic acid and undecanoic acid, which combine with ethanol to form ethyl laurate and ethyl undecanoate, respectively. Isopropyl alcohol presented a similar situation—while not identified in the bourbon or rye whiskeys, the compound is very closely related to isobutanol and isoamyl alcohol, both of which were present in the peated malt whisky and were also identified in the bourbon and rye whiskeys. Again, differences in yeast strain and fermentation conditions can influence the exact profile of fusel alcohols produced from fermentation. Finally, the volatile phenols benzenol (also simply called phenol), 4-ethyl-2-methylphenol, and 4-propylguaiacol likely enter the whisky from the peat smoke when the barley was kilned. It appeared that these esters, fusel alcohols, and volatile phenols unique to the peated malt whisky were just new routes that needed to be added to the already existing ones of my cohesive chemical roadmap.

* * *

However, so far this roadmap was based on flavor compounds *shared* between whiskey and wine. Technically speaking, this map was just a hypothesis. It didn't prove which—if any—of the acetals, aldehydes, esters, fusel alcohols, ketones, lactones, pyrazines, organic acids, terpenes, thioethers, or volatile phenols terroir actually affected.

Luckily, though, we didn't have to stop at the hypothesis stage. After all, Ale and I had already conducted an experiment to probe the effect of terroir in whiskey. I simply needed to overlay our results with the existing chemical roadmap I had developed. If I merged the list of thirty-six flavor compounds (table 9.2) that Ale and I found were influenced by terroir with those compounds that Schieberle, Lahne, and Jeleń had identified in their whiskeys, I could get a comprehensive picture of the chemistry of terroir.

This final overlay revealed that twelve of our thirty-six compounds were identified in one or more of Schieberle's bourbon, Lahne's rye whiskey, and Jeleń's malt whisky. They are noted with the † symbol in table 11.4.

Four of these twelve were correlated with the "good" and "bad" aroma values from our research (table 9.3): isoamyl acetate, ethyl octanoate, nonanal, and 2-nonenal.

Isoamyl acetate was the only one identified across the board, playing a key role in the assessed bourbon, rye whiskeys, and malt whisky. In our study, it was positively correlated with the "good" aroma value. Ten percent of the variation we measured in isoamyl acetate was caused by the farm, and 25 percent of the variation was caused by the interaction of the variety and the farm. Or, said another way, 10 percent of the variation was attributable to the farm environment itself, regardless of variety, and 25 percent of the variation was attributable to how the varieties responded uniquely to the different farm environments. In both cases, genetic and environmental variations led to a distinct presence and concentration of precursor compounds that the yeast would eventually metabolize into isoamyl acetate. Isoamyl acetate is largely recognized as one of the most important flavor contributors in wine, beer, and whiskey. It has a fruity, banana aroma. And, again, it is one of the eighteen flavor compounds responsible for the global vinous odor in wine. It would be a sound hypothesis to say that isoamyl acetate is present in every brand of every whiskey style. So, terroir's effect is on the *concentration* of isoamyl acetate, not the presence. Most importantly, the concentration of isoamyl acetate—and the flavors it imbues—has been proven to directly influence the quality of wine and beer.[7] In wine, more isoamyl acetate typically correlates with quality, complexity, and appreciation.[8] From Schieberle's bourbon aroma recombinate to Ferreira's wine aroma recombinate, the sole omission of isoamyl acetate resulted in a very noticeable deviation from the original sample. It's even been cited as "the only ester capable of imparting its characteristic aroma nuance to wines."[9] Of all the flavor compounds I've identified, discussed, compared, and probed, isoamyl acetate appears to be the most important to consider in the context of terroir.

Ethyl octanoate, 2-nonenal, and nonanal were all identified as key flavor compounds in bourbon, and the first two were included in Schieberle's aroma recombinate. Ethyl octanoate and nonanal positively correlated with the "good" aroma value, while 2-nonenal was filed under "bad." For these three flavor compounds, most of the variation (between 20 percent and 30 percent) was attributable to the corn variety, regardless of the farm. Ethyl octanoate actually possessed the seventh highest OAV in Schieberle's study. In our whiskey terroir research, ethyl octanoate also had the second highest total concentration among all thirty replicates. And like isoamyl acetate, ethyl octanoate is also one of the flavor compounds responsible for the global vinous odor in wine.

Ethyl octanoate, 2-nonenal, and nonanal were identified only in bourbon. This might suggest that their presence correlates with corn in the grain bill, and even then their concentrations are greatly controlled by the chosen variety. So, at least for bourbon or other corn-based whiskeys, these three flavor compounds—along with isoamyl acetate—appear to be among the most important terroir-influenced flavor compounds.

The other eight flavor compounds that we found to be affected by terroir and that were identified in at least one of the assessed bourbon, rye, or malt whiskeys were the esters ethyl acetate, ethyl hexanoate, ethyl undecanoate, and ethyl laurate; the aldehydes 2-heptenal and 2,4-decadienal; the fusel alcohol phenylethyl alcohol; and the acetal 1,1-diethoxyethane. Corroborating our findings, there are ample reports in the research literature (table 11.4) showing that these flavor compounds are also affected by terroir in wine or beer.

There were twenty-four flavor compounds influenced by terroir in our research (table 9.2) that Schieberle, Lane, and Jeleń did not find in the whiskeys they analyzed. We'll look at just three here: ethyl nonanoate, ethyl 4-hexenoate, and styrene.

Forty percent of the variation in ethyl nonanoate was attributable to the farm, 35 percent of the variation in styrene was attributable to the variety, and 30 percent of the variation in ethyl 4-hexenoate was attributable to the interaction. Even though these flavor compounds weren't identified in the whiskey sensomics papers, they could still certainly be flavor contributors in some whiskeys. Their presence in the new-make bourbons we assessed and their absence in the whiskeys assessed in the sensomics papers could be because only specific varieties of corn or specific growing environments encourage their production. Or it could be because of variables beyond terroir, such as yeast strain or fermentation conditions.

For example, while styrene wasn't identified in the sensomics papers, reports have identified styrene as a fermentation byproduct of cinnamic acid metabolism.[10] Cinnamic acid, like other volatile phenol precursors, can be released from lignin and lignan through thermal degradation. At first, that styrene was present in our new-make bourbons but not Schieberle's bourbon was perplexing. But upon further investigation, I realized that it might be because of the yeast strain used by TX Whiskey, which is a proprietary, wild Texas yeast strain. I isolated this particular strain in 2011 from a pecan nut from a ranch in Glen Rose, Texas, on the banks of the Brazos River. The Brazos strain—as we came to call it—always jumped out at me as sharing many of the same flavor characteristics as those created by the Bavarian wheat beer yeast strains. Those same allspice and clove aromas present in hefeweizens (a German word that translates to "yeast wheat") are also present in our whiskeys, especially TX Texas Straight Bourbon. The yeast strains used in hefeweizens are different from most other commercial brewing strains in that they contain enzymes—phenylacrylic decarboxylase (PAD) and ferulic acid decarboxylase (FDC)—that can metabolize ferulic acid to 4-vinylguaiacol via a reaction that removes carbon dioxide (a decarboxylation reaction). The wheat beer strains also employ these same enzymes to convert cinnamic acid (a compound closely related to ferulic acid) to styrene. The presence of these two genes is not the norm for commercial, domesticated whiskey yeast strains.

* * *

As I overlaid the roadmaps, the complexity of terroir grew and grew. It seemed that almost every important grain-derived flavor compound was linked in some way to grain species, grain variety, farm location, or agronomic techniques. Terroir seemed to be more than just some aspect of whiskey production that might influence the presence and concentration of a handful of flavor compounds. The evidence suggested that terroir was one of the most important, all-encompassing factors for grain-derived flavors at large.

But as pioneering and monumental as the whiskey sensomics papers were—and as much as I hoped our research would carry equal weight—there were undeniable shortcomings. It would be shortsighted and outright incorrect to say that the suite of flavor compounds identified in the whiskeys analyzed by TX Whiskey, Schieberle, Lahne, and Jeleń are equally present and important across all brands and styles.

For example, 2-undecanone—the flavor compound I had originally thought was a prime candidate for introducing unique flavors to the new-make bourbons containing it (specifically those from the Mycogen Seed variety grown at Sawyer Farms)—was not detected in any of the whiskey sensomics papers, although other researchers have detected it in brandy.[11] 2-undecanone also did not show any meaningful correlation to the "good" or "bad" aroma values. This fact, however, was largely because many of the batches lacked any concentration of 2-undecanone at all. Given that every batch did of course have some concentration of both "good" and "bad" aroma values, this lack of correlation is to be expected. Other flavor compounds with similar situations (appearing only in a selected few of the thirty total batches) similarly had little statistical chance of possessing a correlation to the "good" or "bad" aroma values.

I went back and assessed the batches made from the Mycogen Seed variety grown at Sawyer Farms that contained a higher concentration of 2-undecanone. They had a distinct pineapple aroma. And while that's a small sample set, it isn't unreasonable to say that 2-undecanone could still be a prime whiskey terroir candidate. It might just be unique to the terroir of Sawyer Farms or to the Mycogen Seed breeding program and their varieties. There are, after all, plenty of examples in wine where one or two compounds are responsible for very distinct flavors (Ferreira's *impact compounds*, again) and where these compounds are present only within specific wine types from particular grape varieties and growing regions. For example, specific methoxypyrazines—one of which (IPMP) possessed a high FD factor in the bourbon assessed by Schieberle—are important specifically in sauvignon blanc, cabernet sauvignon, and Bordeaux wines. And certain volatile thiols, such as 3-mercaptohexan-1-ol (*passion fruit*, *gooseberry*) and 4-methyl-4-mercaptopentan-2-one (*box tree*, *black current*) are most prevalent in sauvignon blanc wines from New Zealand.

The sensory methods used in the bourbon and rye whiskey sensomics papers also raised some concerns. For example, Schieberle claimed that even though acetaldehyde and ethyl acetate contained high FD factors, their inclusion into the aroma recombinate was not necessary. They deemed that their FD factors were not high enough for the corresponding OAV to be greater than 1, and therefore they did not quantify these two flavor compounds. However, Ferreira and colleagues reported that when replicating the aroma of a wine, it was extremely important to include all the compounds with OAVs above 0.5.[12] On the surface, this seems illogical. An OAV below 1 means that the flavor compound is present at a concentration lower than its odor threshold. How, therefore, can such flavor compounds still contribute to flavor? Lahne explains in his thesis:

> There are a number of volatile and non-volatile compounds in whiskey that are not odor-active, but may have a profound effect on aroma synergism, by affecting partitioning or sequestration of actual flavor compounds. It is not just possible but almost certain that this type of effect—of long-chain ethyl esters and other congeners that are not aroma-active per se—is essential to creating authentic whiskey aroma.[13]

So even those flavor compounds that are present at concentrations below their odor threshold can still change the overall taste via synergistic mechanisms, accentuating or suppressing the aroma characteristics and intensities of other flavor compounds.

Schieberle's team only included flavor compounds with OAVs above 1, so it's unlikely that the aroma recombinate was a true flavor clone of the bourbon. In their paper, they claim that "the overall aroma of the bourbon whisky could be mimicked by an aroma recombinate."[14] "Mimic," though, is somewhat of an arbitrary claim, and the sensory panel for the study didn't rate the aroma recombinate as being a perfect match to the bourbon. Further, they used a descriptive sensory technique. While informative, this approach does not provide statistically verifiable results to determine if two products are truly indiscernible. And only orthonasal (sniffing) analysis was employed. The apparent closeness of the two products might have dwindled with the incorporation of retronasal (when flavor compounds escape back into your nasal cavity while tasting) techniques. The retronasal components of tasting are more influential in perceiving flavor than the orthonasal ones.

It's unlikely that acetaldehyde and ethyl acetate had nothing to do with the flavor of the bourbon assessed by Schieberle's team. Both are well-known flavor compounds in all whiskeys, contributing solventy sweet, green apple, astringent notes. Now, it is true that they both have lower boiling points than ethanol and therefore are concentrated in the *heads* portion of the distillation run. Distillers that run pot-still operations will take a *heads cut*, meaning that the first portion of the distillation is kept separate from the *hearts* portion (*hearts* is a synonym for new-make, which is what is actually put into an oak barrel and becomes whiskey). Column-still operations don't allow for such cuts to be made, but there are still techniques that can separate the heads from the hearts. Much of the acetaldehyde and ethyl acetate is sequestered during distillation, but there will always be some of both in the new-make whiskey. It's not uncommon for there to be 10 parts per million (milligrams per liter) of acetaldehyde and 100 parts per million (milligrams per liter) of ethyl acetate in a 100-proof (50% ABV) new-make whiskey, and these concentrations are sufficient for flavor detection. And maturation in an oak barrel only increases these flavor compounds. Whiskey will evaporate during aging, leading to a loss of volume known as "the angel's share." As air enters the headspace vacuum created by the angel's share, it oxidizes some of the whiskey, producing more acetaldehyde and ethyl acetate.

Ultimately, this example with acetaldehyde and ethyl acetate reveals the complexity of flavor, in that flavor compounds below the sensory threshold can still play a role in shaping the overall aroma. It also highlights that flavor chemistry and sensory analysis will always present scientific and technological limitations, and therefore the information is to be taken as a guide as opposed to a definitive code. Flavor is too complex to take a completely reductionist approach to it. There still is—and I believe there always will be—a hint of wonder and mystery to flavor.

Now, to be fair to the Schieberle research group, our whiskey terroir research had some shortcomings as well. For example, we did not identify β-damascenone. Given the importance and universal presence of β-damascenone in wine, that its precursors are universal in grain, and that it was identified as a key flavor compound in the bourbon and rye whiskeys, it's likely that it was indeed present in our new-make bourbon samples (and in the peated malt whisky in the Jeleń paper, for that matter). It's more probable that our SPME extraction method failed to adsorb sufficient levels, or perhaps other compounds masked its presence in the mass spectrometer. Regardless, while we didn't identify it, β-damascenone—along with other norisoprenoid terpenes and their carotenoid precursors—is arguably one of the most important terroir-linked flavor compounds to consider.

* * *

We've covered a lot of scientific data in the past few chapters. But it was important to build a case for the terroir of whiskey through science. Do I think science has answered all of the questions? No. In fact, I doubt it ever will. Flavor is complex and somewhat subjective. But I did discover three takeaways, which I present to you here.

One: We have both experimental and analogical evidence that the terroir of whiskey is real. It changes the chemistry and ultimately the flavor of whiskey by acting on many of the same flavor compounds found in wine. Table 9.2 highlights the experimental evidence. Table 11.4 highlights the analogical evidence, with the column "Influenced by terroir in wine or beer" citing numerous published research articles to support the notion that those compounds most important in whiskey will be influenced by terroir just as they are in wine (and beer). (These sources are found in appendix 3, "Key to the Roadmap: Sources for Chapter 11.")

Two: Even though the overlap was not perfect, the conservation of flavor compounds among the assessed whiskeys and investigated wines meant that a single roadmap was possible. This map (table 11.4, complemented by table 9.2) would be a guide for terroir's effect on the chemistry and flavor of whiskey. There will almost certainly be unique roads specific to styles and brands, and some flavor compounds will undoubtedly be more influential and prone to terroir-driven concentration fluctuations than others. But it appears that corn, rye, and barley (and most likely wheat, by extension) are similar enough physiologically to deliver identical or closely related grain-derived flavor compounds in whiskey. Nuances come from differences in concentration coupled with a smaller number of flavor compounds specific to certain styles and distilleries.

Three: This roadmap was developed by investigating whiskeys made from modern grain varieties, so we can say with confidence that there is still sufficient genetic diversity present among the newer, high-yield varieties for terroir to exist. The genetic diversity is not limited such that all modern varieties of the same species—regardless of environmental forces—deliver the same flavors. But this raises new questions: Does terroir have a more pronounced influence on flavor among varieties that are more genetically diverse, such as heirlooms? If the whiskey industry were to expand beyond commodity grain, using specific varieties, whether heirloom or modern, and highlighting specific farms, might we expand the flavor diversity of the whiskey category that much more?

All important and interesting questions . . . but there was a more immediate issue confronting me. Now that I had it, how exactly was I supposed to *use* this roadmap? How could it guide me to provenance, to the flavor of a place?

Concluding that terroir could still exist among modern varieties (which are not genetically diverse) was an important step because these varieties are still used to make the vast majority of modern whiskey. It also meant that, by default, terroir existed among heirloom varieties (which are genetically diverse). But the anonymity of grain elevators would surely lead to a loss of these terroir-driven flavor nuances. Direct relationships with farmers and directly sourcing grain so it never enters the commodity market—these pursuits would be crucial and necessary for a whiskey to truly express the flavor of a place.

I had experienced the flavor of Sawyer Farms in my whiskeys. But if I wanted to understand the terroir of whiskey fully, I would need to venture beyond my whiskey-making home. I needed to discover who else was sourcing grain directly from farmers, choosing to work with specific varieties, and reintroducing old heirloom varieties. And my journey couldn't live within the walls of distilleries. I would have to travel to the place where whiskey begins—the farm. There were plenty of questions to unravel there. For example, were farmers pursuing their own techniques for flavor? But of all the questions I wanted to answer, one summarized them all.

Who else was out there, trying to chart the same map as me?

PART III

FOLLOWING THE MAP

12

WHISKEY IN THE BIG APPLE

By 2019, I had been a whiskey distiller for nearly a decade. Whiskey occupied nearly every mental faculty I possessed. I was researching it constantly—what had the pioneering scientists and distillers before me already discovered? Who was working on what new projects? What would the next generation of whiskeys taste like?

So when I decided to embark on this journey to discover what terroir meant for whiskey, there were a number of distilleries that I thought would provide the insights I was seeking.

I decided to start with American distilleries, given that they were the most accessible for me. But I also knew American whiskey wouldn't be enough. Whiskey came to the United States with the early immigrants from Ireland and Scotland, two countries still filled with some of the most renowned distilleries in the world. And then there was Canadian whisky and Japanese whisky, what about them? Or the promising—albeit relatively young—distilleries in India, Taiwan, and Tasmania?

Ultimately, though, a complete investigation into every distillery in every corner of the world would be impossible. I had to make a choice, and I chose to limit myself to Ireland, Scotland, Kentucky, and New York. Four headquarters of distinct styles founded on cereal and place: malt whisk(e)y (barley), bourbon whiskey (corn), and rye whiskey (rye).

* * *

I decided to start my journey where the American whiskey industry itself began—the Northeast. From Pennsylvania to Maine, there are dozens of distilleries trying to capture terroir. And the pursuit of terroir starts with local grain.

Some distilleries in New York particularly caught my attention. They were sourcing local grain, yes, but they were also taking it one step further. In 2015, a handful of New York distillers decided to establish a new style of whiskey, which they called Empire rye. To highlight the terroir of their state, they set their own self-enforced rules requiring that 75 percent of the grain bill must be New York rye.

For many years, there had been essentially two grain bills for rye whiskey. Federal law requires that rye whiskey be at least 51 percent rye. Many of the long-standing rye whiskeys out of Kentucky stick to that bare minimum. The rest is usually made up of around 37 percent corn and 12 percent malted barley. You'll see 10 to 12 percent malted barley in many of the grain bills for traditional bourbon and rye whiskey. That's because that amount of malted barley delivers the accepted minimum concentration of enzymes needed to efficiently convert the starch in the rest of the grain bill into sugar. Popular brands like Wild Turkey Straight Rye, Rittenhouse Rye, and Pikesville Rye all use this 51-37-12 formula.

Just across the Ohio River from Kentucky in Lawrenceburg, Indiana, is a company called Midwest Grain Products—MGP for short—formerly Lawrenceburg Distillers Indiana and, before that, the Seagram's Lawrenceburg Plant. There they produce a rye whiskey with a grain bill of 95 percent rye and 5 percent malted barley. Because of the low malt content, they add exogenous enzymes (liquid enzymes produced and isolated from microbial fermentations) to ensure the efficient conversion of starch to sugar.

MGP does not sell a brand under its distillery name. In fact, it doesn't retail bottles of whiskey at all. Instead, it sells barrels to other distilleries, who then blend, barrel age, filter, bottle, and sell it. This is a very common practice called *sourcing*. There is nothing inherently wrong or evil about it, and in most cases, nobody is trying to mislead anyone. Sourcing and blending are how all of the major scotch blended whiskies are produced. There are no Johnnie Walker, Famous Grouse, or Dewar's distilleries. That's because they are not distillers; they are blenders. They source whisky from all over Scotland—of different styles and ages—and blend them to achieve the particular flavor profiles associated with their brands. It's not uncommon for a bottle of blended scotch whisky to contain north of forty different whiskies.

The most well-known brands that have historically sourced MGP's 95-5 rye (as it's called among industry professionals) are Bulleit, Templeton Rye, Angel's Envy Rye, and George Dickel. At the bottling plant, they blend barrels of rye whiskey, each of which possesses distinct nuances, to achieve a uniform product. It's the job of the blenders (who might also carry the title of distiller, especially in the United States) at these distilleries to choose how to marry barrels of MGP's 95-5 rye recipe to create their brand's signature flavor. Some will add an additional step to the process. George Dickel also mellows their rye whiskey in sugar maple charcoal—the traditional Lincoln County process unique to Tennessee distilleries—and Angel's Envy finishes its rye in rum barrels.

Until the emergence of the craft distilling movement of the 2000s, these two rye grain bills—the 51-37-12 grain bill of Kentucky and the 95-5 formula from Indiana—were essentially the only two on the market. And while the whiskey may have been made in Kentucky and Indiana, the rye itself often came from across the country or even across the ocean.[1] Much of the rye used by Kentucky distilleries and by MGP comes from Minnesota, Nebraska, the Dakotas, Canada, or Europe. Back when MGP was still a Seagram's distillery, they started to source much of their rye from Sweden. These Kentucky and Indiana rye whiskeys that use rye from the American Midwest and Europe are high quality and delicious. But the grain flavors in those whiskeys do not capture the provenance of Indiana and Kentucky farms.

And in reality, for more than a generation now, there hasn't been any real opportunity for Kentucky and Indiana distilleries to source local rye.

The commodity grain system defines the market in those states, and so the farmers haven't grown rye. For decades, everyone from seed companies to farmers to distillers have claimed that rye grain fares better in colder climates, which Kentucky and southern Indiana don't provide. So even today, the grain in our Kentucky and Indiana rye whiskeys comes from hundreds and sometimes thousands of miles away.

New York is different. Exactly how cold a place needs to be to grow quality rye grain is still under investigation, but there is no doubt that many of the varieties that were originally brought to America do indeed favor cold-weather conditions. New York certainly has the climate for it, and even though New York is not a huge producer of rye compared to Minnesota, Nebraska, and the Dakotas, it was one of the original farmlands when the grain was first brought over by European colonists. And by extension, rye whiskey—more so than barley, corn, or wheat-based whiskeys—dominated the northeastern whiskey scene in the eighteenth and nineteenth centuries. So when craft distilleries began to emerge in New York, rye whiskey was the obvious choice.

In 2014, after an annual craft distillers' conference, a group of New York distillers were closing out the week with some drinks. Christopher Briar Williams, from Coppersea Distilling, floated the question: What would a New York whiskey look like? It caught fire from there. The founding six distilleries agreed it had to be a rye whiskey, an homage to the state's heritage. I wasn't there, but when I envision this group of pioneering distillers sitting, talking, and drinking around a table while they laid the groundwork for an entirely new style of whiskey—well, it makes me think of the Sons of Liberty plotting revolution in Fraunces Tavern. It's a little exaggerated, I know, but Empire rye was its own sort of revolution in whiskey.

So in May 2019, my wife, Leah, and I headed to New York. I knew the rye was grown in New York State, but I wanted to know more: What varieties of rye? What farms? And what about the flavor chemistry? I wanted to see how Empire rye would fit into my roadmap and if we could uncover any of the unique rye-derived flavor compounds that would differentiate the style from Indiana and Kentucky rye whiskeys.

* * *

My first morning in Manhattan I spent at the headquarters of M. Shanken Communications, the publisher of three preeminent alcohol and tobacco magazines: *Wine Spectator*, *Cigar Aficionado*, and *Whisky Advocate*. Along with our blender Ale Ochoa, we met with a team from *Whisky Advocate* to discuss the goings-on at TX Whiskey, the recent opening of our Whiskey Ranch distillery, and upcoming product launches. And terroir was also a topic of conversation. Terroir was, after all, somewhat of a hot topic in the whiskey world. They were intrigued to hear about our research and findings, and I mentioned that I was in the middle of writing a book. They suggested a few local distilleries to visit, given the work they and I were doing with grain and terroir. A few they mentioned happened to be ones I was already lined up to see over the next two days.

After the meeting—which was accompanied by a few sips of what was at the time our yet-to-be-released TX Texas Straight Bourbon Barrel Proof—I stepped out onto the streets of New York, looking for Leah and her good friend Joanne, a fashion designer who lived in the city. Joanne had promised to guide us via subway to Brooklyn so that I could meet with one of the founding distilleries of the Empire rye movement—the New York Distilling Company. While I might have been able to get myself there—surely missing a stop or a transfer at some point—it was going to be my first subway ride in years, and there was prime shopping on Bedford Avenue, not far from the distillery, that both Leah and Joanne were eager to visit. So I didn't to have to brave the trip alone. (Roll your eyes all you want, New Yorkers. I'm from Texas by way of Kentucky . . . subways are a novelty to me.)

A few weeks before visiting, I had written to the New York Distilling Company, explaining that I was writing a book about whiskey terroir and was interested to learn more about the production of their Ragtime Rye Straight Whiskey. Through research I knew the grain bill was 75 percent rye, 13 percent corn, and 12 percent malted barley and that the rye was grown by Pederson Farms in the Central Finger Lakes region of upstate New York.

One of the founders—Tom Potter—wrote me back and invited me to visit his distillery to talk with him.

I knew the name Tom Potter well. Tom and a brewer named Steve Hindy had founded the Brooklyn Brewery in 1987. Together they wrote the book *Beer School: Bottling Success at the Brooklyn Brewery*, one of the first books I read as I prepared myself to leave graduate school and enter the alcohol industry.

After successfully running Brooklyn Brewery for nearly twenty years, Tom sold his shares and retired in 2004. He planned to spend his time kayaking, writing *Beer School*, and overseeing the American Institute of Wine and Food. A few years later, Potter's wife retired, and they spent a year traveling the world, including time in the South Pacific and at his cabin in Yellowstone. While paddling down America's rivers and traveling the Northwest, he discovered a budding movement, almost like an echo of the craft beer industry. It was the early days of American craft distilling.

In 2008, Tom teamed up with his son Bill Potter and the spirits and cocktail expert Alan Katz to found the New York Distilling Company in Brooklyn. They planned to make whiskey, and they all agreed that that meant rye. They felt that bourbon was firmly associated with Kentucky in the American imagination but that rye whiskey did not yet have a home. It was, of course, already being produced by distilleries in Kentucky and Indiana, but always as a supporting actor to bourbon. Nobody, they felt, had made rye whiskey the star. And they wanted to change that. Katz had been an advocate of rye whiskey for years, and the team was well aware of rye whiskey's preeminence in New York and the Northeast during the eighteenth and nineteenth centuries.

We took the subway to Brooklyn, got off at the Lorimer Street station, and walked north on Lorimer Street. We eventually landed in front of a fire station—Engine 229, Ladder 146. Directly to its right—the brick walls separated by just a few inches—was a squat, one-story steel-façade building, the New York Distilling Company. A slender, gray-haired, middle-aged man emerged from the building to greet me. It was Tom Potter. After a quick introduction, we moved inside, through the bar and distillery, and into a narrow room where Tom shared his desk with whiskey sampling tables and lab benches.

"OK, so the first thing I want to know," I started, "is that while I understand the concept behind local rye, where exactly did the 75 percent of the grain bill requirement come from?"

"Well, actually that was the obvious choice," Tom replied. "In New York, we have this thing called a Class D New York Farm Distiller license. This license—which provides certain tax breaks over a traditional distilled spirits plant license—is only accessible to those New York distillers that use at least 75 percent New York State–grown grain in their recipes."

"Makes sense," I said. "I also read that your rye all comes from Pederson Farms. Who exactly are they? Why did you choose them?"

"They're agricultural pioneers," Tom said. "They were the first to grow commercial hops in New York since Prohibition, planting their first crop in 1999. They are willing to try new things. And growing rye in New York for whiskey was a new thing."

Tom explained that from their earliest days they had sourced rye from Pederson Farms. For years Rick Pederson had grown rye as a cover crop, plowing it back into the earth after growth to protect and fertilize the soil. When Tom first came to Rick about buying rye grain to make whiskey, it was a good prospect for them: it meant their cover crop could bring in some money.

"We didn't pick a particular variety to start," Tom told me. "Rick was already growing a field blend of rye varieties, and that's what he delivered to us. But we were fine with the quality of that rye—it harbored premium and desirable flavors. Plus, I knew that by working with Rick, I might be able to pull off a project I'd been toying with for years."

When Tom opened New York Distilling Company, he envisioned the team creating a style of whiskey that hadn't existed since Prohibition. Tom wanted to create a "whiskey with a story; whiskey that had real roots to the land, and flavor to match."[2] And to truly craft a pre-Prohibition style whiskey, Tom wanted a grain that came from that era. He needed an heirloom variety, ideally one grown in New York centuries ago.

In 2010, Pederson called up the University of Idaho's Foundation Seed Program and asked if they'd send him seeds of their oldest rye varieties. They sent him a dozen seeds of two different varieties—Prolific and Horton, which some plant scientists believe to be indigenous to Eastern Europe and brought to America in the 1700s. These were prime candidates for reviving an age-old flavor and recreating a forgotten whiskey style. But by the time the seeds arrived, the ideal planting season had already passed. The process of increasing seed—especially when starting with only a dozen seeds—can take years before it yields a harvest of sufficient volume to make whiskey.

Not wanting to sit around for a year waiting, Pederson wrote to Mark Sorrels, a professor of plant breeding and genetics and head of the Small Grains Project at Cornell University. He sent Sorrels the seeds, and he grew them in a greenhouse at the Cornell University Agricultural Experiment Station.

"Prolific didn't do well," Potter told me. "Not even in the controlled and ideal environment of a greenhouse. But Horton showed real promise. It was eager to grow."

After that first harvest, it took one more round of seed increases in the greenhouse before there was enough to plant outside. Pederson took the still relatively few seeds and planted them on a four-square-foot plot. This harvest was again replanted, as was the one after that. As was the one after that.

"It wasn't until 2015 that we even had enough to make a test batch," Potter said, laughing. "And then it wasn't until 2017 that we had enough to make an appreciable amount, laying down about a hundred barrels. But I think it was worth the effort. The flavor—it's just so intense. Not necessarily better than the field blend of rye from Pederson Farms we were already using to make Ragtime Rye, but the rye whiskey from Horton has such a concentrated flavor."

"So, a heightened level of flavor as compared to what you make with modern rye varieties?" I asked. (The phrase *heighted level of flavor* is one I've since begun using to concisely explain what I believe local grain and specially selected varieties can bring to whiskey.)

"Yes, definitely," Potter said. "It's been fun. We'll start to release the product in the coming years, probably incorporating the Horton name into the brand. I hope the public appreciates what it is and the work that went into it. It's definitely a gamble—who knows if the average drinker will really care that the whiskey was made from an heirloom variety that took multiple parties more than half a decade to revive. We haven't set a price yet, but it will cost more than our Ragtime Rye made from the field blend. Ultimately, I guess it will just come down to flavor."

"Where, specifically, did the Potter and Horton varieties come from?" I asked.

"It may have originated in Europe and come to America in the 1700s. But there is some evidence that it might have been selected by a family of New York millers with the last name Horton. If so, then it would have been developed in New York sometime in the nineteenth century. We'll probably never really know, but it certainly is a variety whose heyday ended before Prohibition. I'm glad to see it back and excited to see what the whiskey ultimately tastes like after aging a few more years."

I had one more question before our interview ended. I wanted to know whether they had compared the flavor chemistries of their two rye whiskeys—the field blend and the Horton. Both were grown at Pederson Farms, meaning we might be able to isolate a certain aspect of terroir and further understand the distinct flavor chemical profiles from the different rye varieties.

"We haven't done that yet. I think New York State could be in a strong position to explore this type of work: How do different varieties and environments really impact flavor in our whiskeys? What exactly does the whiskey terroir of New York look like? Our state is large, and we have many different climates and soil types. And with Cornell University, we might have the academic collaborator needed to really dive into the science. But for now, we haven't had the time or resources to explore the chemistry. But maybe one day."

This was something of a wake-up call. Cornell University puts New York distillers in a prime position to explore whiskey terroir, but the research is expensive, and universities don't pursue expensive research pro bono. What

Potter said about needing a partner like Cornell made sense—I needed Texas A&M, after all—but it also made me uneasy. This was my first interview since charting the whiskey terroir roadmap. I figured distillers would have information to share and that collectively we'd build upon it. What I was beginning to fear, though, is that the craft distillers on the cutting edge of terroir simply wouldn't have accumulated any chemical data, much less be able to use my roadmap to drive flavor goals.

Here was a central irony. Craft distillers are flexible, eager, and willing to work with farmers by directly sourcing local grain and pursuing wild ideas like reviving forgotten heirlooms. And so much of this fascinating work of capturing terroir was coming from distilleries that then didn't have the money to unravel its mechanisms scientifically.

Does this really matter? Maybe not to most. After all, many craft distilleries are still just trying to break even, pay back their investors, and establish themselves as a viable brand. When a distillery is in this phase, money is usually funneled into sales and marketing, not research and development. But eventually our industry will need to understand the mechanisms behind terroir so that we have the best chance to consistently capture the flavors we seek. And to do this, we need to lay the groundwork with flavor chemistry.

I thanked Potter for his time, bought a bottle of the newly released Ragtime Rye Bottled-in-Bond—made with the field blend rye—and walked down the street to meet my wife at a ramen restaurant for lunch. They didn't sell whiskey there, so I ordered a Brooklyn Lager as a thank you to Tom. Brooklyn Lager is ubiquitous not just in New York but throughout the United States. It's clean and crisp, and its flavors have earned mass appeal. Perhaps one day Empire rye—maybe even one made from the Horton heirloom—will be just as abundant and adored.

* * *

New York Distilling Company was one of six founding distilleries that established the Empire rye style. Another is Tuthilltown Spirits, in the Hudson Valley—equally notable as one of the original craft distilleries, making whiskey as early as 2005. They were also one of the first craft distilleries to be bought; William Grant & Sons acquired them in 2017. The other four distilleries were Finger Lakes Distilling, in Seneca Lake (the same location as Pederson Farms); Coppersea Distilling, in Ulster County; Black Button Distilling, in Rochester; and Kings County Distilling, in Brooklyn. Kings County has made whiskey there since 2010, making them the oldest distiller in New York City since Prohibition.

I first learned of Kings County Distillery in 2010 when I was still in graduate school at UT Southwestern Medical Center. I was always curious about their distillery, even though I had never even tasted their whiskey before this trip to New York. One of their founders—Colin Spoelman—was originally from Kentucky. So while I was considering where and how to join the whiskey industry, Colin provided some inspiration. If he could leave Kentucky to make whiskey in New York, then maybe I could do the same in Texas.

Further, the original setup at Kings County made the notion of starting a distillery less daunting. For their first couple of years, they were an almost impossibly small operation. Their distillation system was essentially a home-hobbyist setup: five-gallon stainless-steel pot cookers, plastic fermenting buckets, pot stills that looked as though they were hammered out of old metal milk jugs, and oak barrels that could hold only a couple of gallons. Because of these small barrels, they bottled their whiskey after only a year or two of maturation.

I should note that the use of small barrels doesn't allow one to age whiskey faster. A whiskey aged for one year in a five-gallon barrel will not taste like a whiskey that's been aged for four years in a standard fifty-three-gallon barrel. Small barrels expedite the extraction of wood-derived flavor- and color-inducing compounds, but there is no substitute for time when it comes to the oxidation reactions that are equally important for whiskey maturation. These oxidation reactions occur as air diffuses through the wood, and they help soften tannins, convert soapy and fatty aldehydes to floral and fruity notes, and facilitate the creation of acetal, an important contributor to the bright top notes of well-matured whiskeys. This means that whiskey aged in a small barrel might need to be removed from the oak—lest it become overly woody—before oxidation has balanced the flavors. If so, this can lead to a somewhat one-sided and oak-forward whiskey. But it doesn't mean small barrels necessarily make *bad* whiskey. The flavors are just different, and the distiller has to be more concerned with overly oaked flavors than he or she would when using larger barrels. Kings County Distillery is an example of a distillery that can produce high-quality, award-winning whiskey in smaller barrels. So when it came time to pick New York distilleries to visit and investigate, Kings County was at the top of my list.

I visited Kings County Distillery the day after my meeting with Tom at New York Distilling Company. They had come a long way from when I first discovered them. For one, they had moved. What started in an apartment-sized, 325-square-foot room in East Williamsburg had moved into the Paymaster Building in the Brooklyn Navy Yard in 2012. Their second home is "just steps away from the legendary site of the Brooklyn Whiskey Wars of the 1860s."[3] They also now reside in the former distillery district for the Brooklyn waterfront. They've grown far beyond five-gallon buckets and milk-jug pot stills. They now have over thirty thousand gallons' worth of fermenters, a Vendome five-thousand-liter pot stripping still, and two Forsyths copper-pot spirit stills with a total capacity of 1,650 liters. Forsyths is the preeminent manufacturer of distillation equipment in Scotland, and their U.S. equivalent is Vendome Copper & Brass, based in Louisville, Kentucky. The stripping still conducts the first distillation of beer into low wines, which are then portioned out and distilled again in the spirit stills into high wines, a synonym for new-make whiskey. Kings County Distillery now ages whiskey primarily in fifteen- and thirty-gallon barrels.

Walking up to the Brooklyn Navy Yard, you are greeted by two brick buildings bisected by a gate. Both buildings possess towers and castle façades, giving you the feeling you're about to cross a drawbridge and gatehouse. The right building had a window labeled "Tasting Room," while the left building was labeled "Office and Lab." I headed for the Tasting Room.

Inside we found a nearly empty room, with a lone employee behind a coffee bar. Kings County Distillery did more than whiskey, it appeared.

Upon telling the employee that I was here to visit the distillery, they directed me to head out the building, turn right, and walk down the street a few hundred yards. Nestled among the operations of what is still an active shipyard, we found a two-story, red-brick building with a stone sign that read "Paymaster of the Navy Yard." New York's oldest distillery since Prohibition would be inside.

We met with Gabriella Gjonaj, the chief operating officer. Kings County is separated by two floors—the top holds the offices, visitor center, and a site for barrel maturation, and the bottom is the distillery itself. Gabby walked me around the distillery, and we tasted a selection of their whiskeys.

"You make a lot of corn-based and rye-based whiskeys here," I said to Gabby, as we discussed their current whiskey offerings. "Tell me about the grains you use to make them."

"The corn we use is all organic, and it comes from Lakeview Organic Grain in the Finger Lakes region."

Lakeview Organic Grain is a farming operation owned and operated by Klaas and Mary-Howell Martens, whom I had previously read about in Dan Barber's book *The Third Plate*. Barber attributes his shift in working with generic white flour to working with flavorful, organic whole grains to Klaas and Mary-Howell and the wheat they grow. Along with the farm, the Martens also own a grain elevator and mill operation, but they buy only organic grain from farmers in the Finger Lakes region.

The Martens farm corn, soybeans, flax, barley, spelt, oats, and wheat. Of corn, they primarily grow organic hybrid varieties but also some heirlooms. An organic hybrid corn variety will look identical to a nonorganic one. In fact, a variety in and of itself cannot be organic, because "organic" only refers to how it is farmed and whether it has been sprayed with chemical fertilizers and pesticides. Or at least that was the case before genetically modified organisms. If a variety contains GMO traits, by default it cannot qualify as organic, even if it is farmed to organic standards. Regardless, almost every hybrid corn grown today classifies as a modern variety, and they all are genetically very similar. According to Klaas, "The vast majority of corn varietals are no longer being planted at all. The corns people grow today are such a narrow piece of what's out there historically."[4]

Still, the Martens attribute the quality of their hybrid corn to the organic farming methods they utilize. "Organically grown hybrid corn, if it's grown in healthy soil anyway," says Mary-Howell, "is of a much higher quality than the same variety grown conventionally. It's of a higher quality, and it has more flavor and nutritional content when it's grown in healthy, nutrient-rich soil. Now, not all organic corn is grown in healthy, nutrient-rich soil, and if that's the case, it's not going to be more flavorful or nutritious, but that's how we do things here. Everything we do starts with healthy, living soil."[5]

What she's saying is that if the same two varieties are grown on two fields of the same farm (I'm stretching her words a bit with the "same farm" point, but I think Mary-Howell would agree), where one field is farmed organically and the other conventionally, the corn from the organic field would taste better, that is, have what I would call a "heightened level of flavor." In essence, that's one *farming-for-flavor* element of terroir.

"The organic system, in order for it to work, is based on intentional biodiversity," says Mary. "It's based on having a large variety of plants in the fields. Conventional farming is just the opposite—it's all about monoculture and intensive, large-scale production of one crop in one place. For us, the greater the number of species we have on our farm, the better, because they're going to form a system that's in balance, that can sustain and nourish and protect itself through its own diversity, without needing intervention in the form of chemical fertilizers, herbicides, or pesticides. That system includes both animals and plants, and it includes the animals you see and the animals you don't."[6]

What Mary was referring to were microbes and the importance they play in soil fertility.

"One key," Mary continues, "to giving all of those little animals that create such healthy soil the conditions they need to thrive is crop rotation. We're always rotating different crops, different varieties of a crop, and different cover crops on our fields from season to season, so we're constantly replenishing the nutrients in the soil. Healthy, living soil allows the plants growing in it to be much healthier and more resilient. They're better able to compete against weeds and pests, and they're going to be healthier and more flavorful than conventionally grown versions of the same plants."[7]

I next asked Gabby about the rye they use to make King County's Empire rye whiskey.

"All of the rye for our Empire rye whiskey comes from Hudson Valley Hops & Grains, in the Hudson Valley," she said, "and the variety we use is Danko. Our grain bill is 80 percent rye, 20 percent malted barley."

Danko was not the same field variety that Pederson Farms grew for New York Distilling Company's Ragtime Rye. And Hudson Valley Hops & Grains was over two hundred miles east of Pederson Farms, far enough of a distance to register as an environmentally distinct region. This meant that the two whiskeys might be ideal for a comparison of how terroir can affect flavor in two Empire rye whiskeys. Both were distilled and aged in the same borough of New York City, removing the very influential variable of maturation location and environment. There was a very small divergence in grain bill, and the yeast strain, distillation technique, and size of the barrel would all influence flavor as well—and therefore must be taken into account—but it was still likely that an important flavor difference between the two Empire rye whiskeys would indeed be attributable to the place where they were grown. Or, more specifically, how two individual varieties of rye expressed flavor through two distinct growing environments.

Aside from differences in variety and environment, as I read through the websites of Pederson Farms, Hudson Valley Hops & Grains, and Lakeview Organic Grain, I became equally interested in the potential divergences of farming techniques. Pederson Farms grew both organic and conventional rye. Lakeview Organic Grain practiced only organic farming. And Hudson Valley Hops & Grains claimed that they pursued a mingling of organic and biodynamic techniques. I wanted to understand the differences—specific to these three farms but also in a more general context—among these farming techniques. Not so much to learn how they can differently affect phenomena such as environmental sustainability and human health: topics such as these warrant their own books, and many have indeed been written. Instead, I wanted to see what the science said about how the flavor of a crop can be influenced by different farming techniques.

13

THE TRILOGY OF FARMING

WHILE DIFFERENT TERMS are sometimes used, there are three basic types of agricultural management practices: conventional (sometimes called industrial), organic (sometimes called biological), and biodynamic. The dominant agricultural practice today is conventional farming. These farms typically focus on monocultures of high-yielding crop varieties. Monoculture is the practice of growing a single crop species or variety in the same field at any one time. The benefit is efficiency and output. The risk is that disease, drought, or some other catastrophe can destroy an entire crop at once.

Monoculture is also found in organic farming. So what separates conventional and organic farming? One difference is the presence of genetically modified crops (often called GMOs). Organic farming does not allow GMOs, but they dominate the corn, soybean, and cotton fields on conventional farms. As of yet there aren't any GMO wine grapes, so in winemaking the main differentiator is the use of synthetic fertilizers, herbicides, insecticides, fungicides, and other pesticides. Conventional farming embraces them—sometimes truly relies on them—for agronomic success, but organic farming prohibits most of them. Because conventional farming can deplete the soil, it requires heavy synthetic inputs to grow crops successfully.

Organic farming, on the other hand, relies on other methods to maintain soil health and ensure a bountiful yield. One method is called *low tillage* or even *no tillage*. Tilling is like plowing, except tilling runs blades through the soil, whereas plowing digs into the soil and flips it. Tilling loosens the soil, which makes planting easier. It also destroys weeds and mixes leftover organic material (like stalks left after harvest) back into the soil. But it can also disrupt the soil's organic matter and microbial ecosystem, which are the twin hearts of soil health. Tilling also encourages soil erosion. To avoid tilling, new planting technology drills seeds directly into the soil without disturbing the leftover organic matter from the last harvest, reducing erosion and improving water retention. It's important to note, however, that many organic farmers still do indeed till in order to kill weeds (as they are not able to use chemical herbicides) and to plow cover crops back into the soil as organic matter (as they are not able to use chemical fertilizers).

Another method organic agriculture uses to ensure soil health and yield is to lean heavily on crop rotation. This is when two crops are planted in successive seasons on the same field. The two crops usually have a sort of mutualistic relationship. A common pairing is cereal grasses (grains) and legumes, since the legumes will fix nitrogen from the air into the soil, creating suitable conditions for growing grains, which do not fix nitrogen and must instead absorb it from the soil. The benefit of crop rotation is that both crops can be harvested for use or sale. That's not the case with cover cropping, another method employed in organic farming to maintain soil health. Cover crops are grown but then plowed back (in tilling operations) or rolled (in no-till operations) into the ground rather than harvested, replenishing the soil with organic matter and building a sturdier, erosion-resistant soil.

Crop rotation doesn't work with grapevines, because they are not annual plants like grains and legumes. Some vines are over a hundred years old. But cover cropping is possible, and some vintners will plant yellow mustard and clover around grapevines to protect the soil and minimize insects and other pests.

Biodynamic agriculture is a close cousin to organic farming. Biodynamic farmers even adhere to the same restrictions as organic farmers. The difference between the two is that biodynamic farms operate as a closed system, meaning that any and all inputs used to grow a crop must also be produced by the farm. For instance, manure and compost are powerful natural fertilizers. But whereas an organic farm might import them from outside, a biodynamic farm would produce them itself. This would mean the biodynamic farm would also need to raise livestock (for manure) and maintain a composting station. And in turn, if livestock are being raised, then the biodynamic farm would also need to grow grains or other grasses for animal feed. Biodynamic farmers also work with specific fertilizers (called "biodynamic preparations"), which are largely absent from organic farming.

To its proponents, biodynamic agriculture is the pinnacle of humanity harmonizing agriculture with nature. Achieving this balance means that the operation must be extremely diverse in the crops grown and the livestock raised. In this way, biodynamic farming is the antithesis of monoculture farming. But harmony and diversity come at a cost. Biodynamic farming is inefficient and labor intensive compared to conventional and even organic (at least large-scale organic) farming.

On this spectrum of efficiency to harmony, conventional agriculture requires the most inputs, the majority of which are synthetic chemicals. Organic agriculture typically—though not always—relies more heavily on crop rotation and cover cropping and therefore needs fewer inputs. Those inputs must be "natural," like manure, compost, and vinegar. But they often still come from beyond the farm. Biodynamic farming means using only inputs produced on the farm itself.

A trouble in terms is that *conventional farming* is a blanket term covering everyone from farmers who skillfully incorporate standard methods of organic and biodynamic farming to those who follow every worst practice of industrial agriculture. There are many conventional farmers who drill rather than till, rotate crops and grow cover crops, and spray minimal or no pesticides. A farmer could even follow almost every rule for USDA Organic certification, but if he or she plants a GMO crop or sprays even a tiny amount of synthetic pesticide to protect a cropping or rotation, then it is still classified as conventional agriculture. This is why "conventional" is often split into two subcategories—low input and high input. Sawyer Farms, which supplies TX Whiskey with the grains we use to make our whiskeys, is an example of an ecologically conscious, sustainable, low-input—but still conventional—farm.

Aside from the potential flavor differences these farming techniques might cultivate, they raise another question: Is terroir most effectively expressed when crops are farmed to balance their growth with the environment, rather than being strained and stretched by industrial farming practices? You can force grains—especially the modern, high-yielding, commodity grains—to grow on almost any plot of land with the right inputs. But to what extent do grains grown in this manner really express terroir? Instead of possessing the flavor of a place, would these grains not instead express the extrinsic and standardized characteristics produced by industrial agronomic techniques?

I think of it like this: many animals can be held in captivity, but some of them do not *thrive* there. No one as yet has kept a great white shark—the most powerful killing and eating machine of our time—alive in an aquarium. They simply refuse to eat and so starve to death. Why they do this is not entirely known, but for an apex predator to refuse a meal to the point of starvation shows how profound the changes can be when a species is out of place in an artificial or forced environment. The plant or animal is no longer an expression of its true self or of the environment that selected for it. Instead, it has become an expression of human intervention.

I had already accepted that terroir was more than just the environment. It was the overall ecological system of a farm or vineyard, that complicated interplay of nature and nurture. Now the question was whether terroir's expression required that ecology to be *balanced*, that is, an ecology where Mother Nature and human intervention were in harmony.

* * *

Unfortunately, there were no studies on the effect of farming techniques on flavor in whiskey. But once again, the wine industry came through.

According to John Williams, the winemaker at Frog's Leap Winery in Napa Valley, "Organic growing is the only path of grape growing that leads to optimum quality and expression of the land."[1] Ron Laughton—the owner of Jasper Hill Vineyards—elaborates on this idea:

> Flavors are created in the vine. The building blocks are the minerals in the soil. If you keep applying synthetic chemicals, you are upsetting the minerals in the soil. So if you wish to express true terroir, you should be trying to keep the soil healthy. Let the minerals that are already there express themselves in the flavor in the vine. Herbicides upset the balance of the vineyard simply because dead grasses are an essential part of the vineyard floor. Those dying grasses act as food for another species, and [those species in turn] act as food for another species. You go right down the food chain to the organisms that create the minerals for your plant to suck up and create the building blocks for the flavors. It's not rocket science.[2]

In a 2014 survey of over three hundred owners and managers of California wineries conducted by UCLA, 25 percent named "improved quality of grapes/wines" as their top motivation for adopting sustainable, low-input techniques.[3] High-input conventional practices have a tendency to reduce soil microbes, as over-tilling and the heavy use of synthetic chemicals can decimate their soil, vine, and grape environments, killing off microbial life and throwing the ecosystem out of balance. Recent studies have found that the microbial life in vineyard soil, on the vine, and on the grapes themselves can change the flavor of the wine.[4] Given that the indigenous microbes (especially lactic acid bacteria) that enter the whiskey-making process through grain and malt do help develop flavor through late lactic fermentations, the same phenomena might also change how the flavor of terroir is expressed in whiskey.

Then there are the physiological and biochemical aspects of the grape itself. Different agronomic techniques have been shown to affect the balance of reproductive growth (fruit yield) and vegetative growth (pruning weight). An unbalanced vine can produce lower-quality fruit and impede the concentration of sugar, phenols, and anthocyanins.[5]

This all points toward agronomic techniques influencing—possibly substantially—grape and wine flavor, but I still lacked concrete evidence. If I could discover what the wine science literature that used analytical chemistry and sensory analysis had to say on this topic, then maybe I could determine with real confidence whether the flavor of a wine is influenced by the type of grape—conventional, organic, or biodynamic—from which it is made.

In 2004, Jean Reilly reported a flavor comparison of biodynamic and conventional wines.[6] The organizer gathered ten conventional and ten biodynamic wines. Pairs of each were matched—based on how close the two vineyards were, the vintage, and the price range—and then presented to wine industry professionals in a blind taste test. In four out of five cases, the professionals preferred the biodynamic wine over its conventional counterpart. The study was published in the magazine *Fortune*, which, while full of high-quality and vetted content, is not a peer-reviewed scientific journal, and no other scientists monitored the investigator's experimental methodology. This doesn't necessarily mean the data are wrong or uninformative, but they should be taken with a grain (pun intended) of salt.

That said, there have been plenty of academic articles that have taken similar approaches. The difference is that these academic articles utilized a much larger sample size. Consider the 2016 statistical study by researchers from the University of California–Los Angeles and the KEDGE Business School in Bordeaux, France, published in the *Journal of Wine Economics*.[7] The researchers did not conduct an actual experiment here. Instead, they collected data from a wide range of "experts" and used statistical methods to draw inferences about the differences in flavor and quality between conventional wine and organic/biodynamic wine. The experts were not trained sensory professionals at a university, and the data were not derived from any scientifically vetted sensory method. Instead, the data were gathered from three popular-press publications—*Wine Advocate*, *Wine Spectator*, and *Wine Enthusiast*—each of which assesses wines using hundred-point scales developed internally and without any peer review. And in reality, these "hundred-point" scales start at fifty. Aside from the tastings usually being blind, there is not much about this rating method that is any more scientific than the *Fortune* study.

But there are still advantages to using such data. Most importantly, the number of samples in the study was far beyond what is feasible for a single scientific study. A sensory panel might accomplish dozens or—for extremely well-funded experiments—hundreds of nosings and tastings, but this *Journal of Wine Economics* study incorporated over 74,000 tastings of different California wines over all vintages between 1998 to 2009. And even though these expert raters are not scientists, they are almost surely more experienced with the spectrum of quality that can exist among wines than any trained sensory panel. Their job, after all, is not to deconstruct a wine for a scientific study but to unravel its flavors for readers and drinkers. In one sense, this reduces the scientific validity of their ratings. But in another sense, it increases its practicality. It might be a much better indicator of the all-important questions: What does this wine taste like, and how good is it? Ultimately, the scientists reported that among the more than 74,000 California wines included in their statistical analysis, organic/biodynamic wines enjoyed significantly higher ratings than conventional wines.

Insightful as these results were, the researchers had grouped organic and biodynamic wines into one category—ecocertified. The first report (at least in a peer-reviewed journal) that tested if wine flavor was influenced by biodynamic versus organic techniques came out in 2009. Published in the *Journal of Wine Research*, researchers at Washington State University grew merlot grapes on a twelve-acre vineyard, separating plots in a randomized order based on whether they received biodynamic or organic treatments.[8] The main difference between the two treatments was that the biodynamic plots received the fertilizer preparations that are specific to their technique. They repeated this growing regime over the course of four vintages (2001 to 2004). Ultimately, for each vintage, they detected no major flavor differences between the organic merlot and the biodynamic merlot. The 2004 vintage was the exception: the organic merlot had notably higher astringency and bitterness and a longer finish than the biodynamic merlot. But even here, the difference was not statistically large, and overall, the experiment suggested that the two approaches did not lead to drastically different wines.

But for most of the scientific literature I found—whether the comparison was between conventional and organic/biodynamic or between organic and biodynamic—the data were conflicting. Some reports showed that agronomic technique influenced flavor; others claimed no significant differences. But in February 2019 came my lucky break: the *American Journal of Enology and Viticulture* published an extensive review that aimed "to review evidence comparing effects of conventional, organic, and biodynamic viticulture on soil properties, biodiversity, vine growth and yield, disease incidence, grape composition, sensory characteristics, and wine quality."[9] The abstract of the paper concluded by saying, "By describing different hypotheses concerning the effects of organic and biodynamic viticulture, this review and meta-analysis provides helpful guidance for defining further research in organic agriculture on perennial, but also on annual crops."

It was as if the researchers were saying to me, "We know whiskey distillers are curious about this too. So, take notice: Our findings matter to those annual grain crops you use." But my luck was short-lived, as the review paper quickly confirmed what I had already discovered through my own investigations in the literature: the effect of agronomic technique on the flavor of wine is inconclusive. In some studies, wines produced from organic and biodynamic grapes were found to have distinct—oftentimes preferred—flavors from conventional ones. But other studies found no difference or even found wines produced from conventional grapes to be more complex.

Similar to other variables explored in this book, it may be difficult for science to definitively answer the question whether organic and biodynamic agricultures create crops with premium flavor. Flavor is a complex phenotype, and its experience is to some extent subjective to the taster. It doesn't always lend itself well to scientific analysis. Further, all the processing steps between harvest and the food or beverage landing on your plate or in your glass introduce instances when flavor can be influenced by something else. The answer may always be a mixed bag of yes, no, and maybe, contingent on any and all phenomena associated with the study or tasting. That said, there is at least *some* evidence—from the popular press to scientific journals to the opinions of chefs and expert tasters—that organic, biodynamic, and low-input agricultures can produce distinct, diverse, and local flavors better than high-input, high-yield conventional farming can. Just as much as high-yielding varieties and grain elevators, high-input conventional agriculture has aided the commoditization of grain.

A closing note on this background into farming and flavor: It may seem that my main interest in different agronomic techniques is less about the environmental impact and more about producing great flavor. While this is true in the specific context of this book, the effects of high-input conventional farming on our global ecosystem can't be ignored. That said, environmental consciousness and flavor consciousness might work in a virtuous circle. In *The Third Plate*, Dan Barber, the chef at the helm of Blue Hill at Stone Barns, wrote, "When you pursue great flavor, you also pursue great ecology."

Because synthetic chemical fertilizers and pesticides aren't there to mask or supplant it, low-input, ecologically friendly agriculture should express the essence of an environment through its crops. Even if terroir-driven local flavor is the only goal, it is inseparable from the effect on—or, perhaps more correctly, the interaction with—the environment. So I would say, "When you pursue terroir, you pursue great flavor, and with it, great ecology."

* * *

A few months after my visit to New York, I called Mary-Howell Martens at Lakeview Organic Grain. I brought up what Dan Barber had written and asked her to weigh in on how her farm targets flavor through her farming methods. I opened the conversation by describing the book I was writing, but Mary-Howell stopped me midsentence. "As a scientist," she said, "I don't want to perpetuate the idea that we can taste terroir in whiskey."

Tough crowd, I thought.

Along with owning and operating farming and grain elevator ventures, Mary-Howell has a master's degree in plant breeding from Cornell

University. And as a scientist, she realized—rightfully so—that terroir was just as easy to exploit as it was hard to prove.

"There's just so much subjectivity in flavor," she told me.

I had read an interview with Mary-Howell in which she tells the story of an organic grower in New York who put out two piles of corn ears in his barn. One pile was organic corn, and the other was conventional. At night, the barn mice came out and feasted upon the organic corn ears and ignored the conventional corn. They did this night after night, even when he switched the location of the piles. But while such anecdotal evidence is revealing, and even though she is an ardent supporter of organic agriculture, Mary-Howell says there is still very little scientific evidence to support any definitive claims on better flavor.

I assured Mary-Howell that while terroir was sometimes wielded as a meaningless marketing buzzword, my goal was to lay some scientific groundwork for its validity and also define the boundaries of its influence. From her tone, I could tell she was still suspicious, but I pressed on, asking her to discuss the differences in techniques and why organic and biodynamic farming might produce better-tasting food than conventional (or at least high-input conventional) farming.

"Healthy soil makes a huge difference," she started, "or, more specifically, the organic matter in the soil. Chemicals from conventional agriculture can kill the microbial life in the soil."

If organic and biodynamic techniques lead to more flavorful food, it might largely be because of the microbial life thriving in their soils that cannot survive in the soil of high-input conventional farms. A diverse microbial soil community produces a set of molecular nutrients in the soil, which provide plants with a sufficient and diverse set of chemical building blocks and stimulate chemical production in plants.

I asked Mary-Howell about heirlooms, and she seemed more optimistic about their potential for discovering new flavors than she was about the effects of farming technique. "Heirlooms have more complex flavors," she told me, "but customers are not always quick to commit to them. They are initially excited about the prospect, but given they yield about half as much as hybrids, they cost twice as much. That cost increase is not something most customers are interested in."

Last, I asked her whether there were any techniques used to grow grain that were specifically employed for flavor, similar to how vineyard managers will employ techniques to target specific flavors in their grapes. "Not that I know of," she replied, "and chances are that such techniques would lead to lower yields and therefore cost the customer more. Farmers would explore such techniques if there was a commitment to quality—and increased cost—from the customer. As of now, I don't see that commitment."

What Mary-Howell and Klaas are doing on their farm and at Lakeview Organic Grain can't be summed up as a philosophy bent toward the yield as the "almighty goal." Now of course, yield is indeed *one* goal, but fostering a healthy environment is another, and cultivating nutritious, delicious food is a third. While their farming methods were not ones they specifically pursued for flavor, they were geared toward an overall healthy ecosystem, and one of the rewards for doing so—whether intended or not—was flavor. While it wasn't the same as how vineyards are managed for flavor—which may ultimately never be the case for grain farming—their efforts were a step in the right direction. And while it is difficult to prove scientifically, the whiskeys produced from their grain, at least to some extent, will harbor nuances distinct to their farm and their approach to farming.

* * *

The output of Empire rye among all New York distillers is still tiny compared to the rye output from Kentucky and the brands that source Indiana whiskey from MGP. And honestly, for most of my trip to New York City, the presence of Empire rye was hard to detect, much less taste. Maybe that's a sign of its youth or simply of the massiveness of the city. When I asked the waitress at our Times Square hotel if they had any Empire rye whiskey, she gave me a confused stare. When I rephrased my question, asking if they had any whiskey from New York, she simply pointed me to the list of whiskeys in their cocktail book—and Empire ryes weren't there. The style is still incredibly young, and I believe its potential is far from realized.

I think too that the work being done can serve as inspiration to other states and distilleries. I was greatly inspired by Tom Potter's success in reviving an heirloom rye variety, so much so that a similar project has become the second phase of my PhD research. My goal is to revive an old Texas heirloom corn variety called Hillsboro Blue & White. This variety has been tucked away in the seed vault of Seth Murray (my PhD advisor) and his predecessors for decades now. While little is known of this variety, the name suggests that it was selected in Hillsboro, Texas, which is where Sawyer Farms is located. We hope to not just increase the small amount of Hillsboro Blue & White seed that is left but also to employ plant breeding selection techniques. Our aim is to further adapt it to the environment of Sawyer Farms while concurrently tailoring it for whiskey making. One whiskey-making trait we will select for is alcohol yield, but not at the expense of flavor. Flavor will be our primary selection trait.

While the reviving of heirlooms is very interesting, Empire Rye's most meaningful inspiration will most likely come from its requirement for locally grown grain. It is the first whiskey style regulation—whether enshrined in law or simply self-policed by distillers—that echoes the regulations of wine terroirs, which require the majority of the grapes to be grown in the respective appellation or region before it earns the geographic distinction on its label. And Empire rye has inspired the rise of other new styles that also mandate local grain. On July 11, 2019, the Missouri state government signed a bill mandating that for a bourbon to be labeled as "Missouri bourbon," all of the corn in it must have been grown in the state. Until now, most whiskey style regulations only require that the whiskey is distilled and aged in the state or region on its label.

But if a whiskey produced in Texas, for example, used barley grown and malted in Scotland or heirloom corn that comes from Iowa, is it truly a Texas whiskey? Should not the farm and the grain be the launch pad from which regional distinctions and flavor nuances are created? While many craft distilleries in America source local grain, there is actually another place that many might be surprised to find out is the leader in this movement, given its historical reliance on imported commodity grain. But this place might have the funds, resources, heritage, manpower, and pioneering spirit to make local grain more than an ingredient found only in niche whiskeys. This place has the potential to influence state and federal regulations and change the landscape of American whiskey, bringing local grain (not to mention sustainable farming and agroeconomic development) back to the forefront.

This place is Kentucky.

14

MY OLD KENTUCKY HOME

No state in America has a more storied or respected tradition of whiskey making than Kentucky. There are over 9 million barrels aging in the Commonwealth—two for every Kentucky resident. Its agricultural industry has an equally storied history. Kentucky's most important crops are tobacco, soybeans, hay, wheat, and corn. The latter two (especially corn) are important ingredients in many Kentucky whiskeys (especially bourbon). By law, bourbon must contain at least 51 percent corn. Traditionally, either rye or wheat is the small-grain complement to corn in a bourbon grain bill. Rye is the more common choice—Jim Beam, Wild Turkey, Woodford Reserve, Buffalo Trace, and Evan Williams all use rye in their bourbon. When a distiller uses wheat, the spirit is called a *wheated bourbon*. Maker's Mark, Pappy Van Winkle, and W. L. Weller are some of the better-known wheated bourbons. Both approaches make fantastic bourbons. Bourbons that complement corn with rye tend to be spicier. Those that complement corn with wheat tend to be softer and a bit sweeter.

Given the importance of corn and wheat in whiskey, and given that these are principal Kentucky crops, you wouldn't be faulted for assuming that all Kentucky wheated bourbons would be produced from Kentucky-grown corn and wheat. But this is not the case. Up until 2014, only about

40 percent of the corn used to make Kentucky bourbon actually came from Kentucky.[1] The vast majority of the remaining 60 percent came largely from grain elevators in Indiana. These percentages have recently shifted in favor of Kentucky corn, largely thanks to lobbying by the Kentucky Corn Growers Association. But even today, much of the corn used by Kentucky distilleries is grown to the north: in Indiana, Illinois, and throughout the Corn Belt.

Regardless of where the corn, wheat, or rye comes from, no other state uses more commodity grain to make whiskey than Kentucky. Kentucky distilleries will usually not specify grain characteristics beyond the following: yellow dent corn, soft red winter wheat, plump rye, and malted barley, all of which must meet acceptable USDA grades, usually #1. So most Kentucky whiskeys are produced from commodity grains, and many of these commodity grains are grown somewhere else.

On the surface, a good explanation for the use of commodity grain would be the overwhelming production volumes of the largest Kentucky distilleries. On the whole, the Kentucky bourbon industry consumes 15 to 20 million bushels of corn per year.[2] In 2019, Kentucky growers averaged about 180 bushels per acre.[3] So an estimated 83,000 to 111,000 acres of Kentucky farmland would be needed to supply the Kentucky bourbon industry. This might seem like a lot of land, but consider that more than half of Kentucky's total 25.4 million acres is devoted to farmland and that 1.5 million of those farming acres are already devoted to corn. Redirecting just 5 to 8 percent of all Kentucky corn growing could supply the entire Kentucky bourbon industry.

As a specific example, Jim Beam—one of the largest Kentucky bourbon distilleries—produces somewhere in the neighborhood of 500,000 barrels per year.[4] On average, eight to ten bushels of corn are needed to produce one barrel of bourbon. Jim Beam will consume anywhere from 4 million to 5 million bushels of corn each year. At 180 bushels per acre, 22,000 to 28,000 acres of farmland would be required to supply Jim Beam with their annual corn needs. This may seem like a lot of farmland. But consider that the large wine operations have similar acreage requirements. E&J Gallo, for example, owns over 23,000 acres of vineyards across the state of California.[5]

Further, they maintain contracts with growers throughout the state to bolster their yearly needs.

Why did the large Kentucky distilleries continue to steer away from the potential of terroir—and local grain—over the last hundred years, locking themselves into the commodity grain market? There are a number of reasons—financial, agricultural, operational—but perhaps more than anything, it was because there was no substantial evidence that selecting and controlling for terroir in grain would pay off.

That said, even in Kentucky, some researchers and distillers have finally turned to terroir. Some of the most exciting work being done to explore the influence of grain variety and farm environment on flavor is coming from the big bourbon distilleries. So I decided it was time to see some of these efforts firsthand.

* * *

A month after our trip to New York, Leah and I headed to my hometown, Louisville, Kentucky. The city would be our home base for a multiday excursion in search of terroir through bourbon country. Growing up in Louisville, I had already visited many of the Kentucky bourbon distilleries. I knew the methods and ingredients they used and what they do and do not communicate to guests on their whiskey tours. Take any of the dozens of distillery tours on offer in Kentucky, and in each one you will hear about three things.

The first is the water. Kentucky's aquifers all sit on beds of limestone, which is a carbonate sedimentary rock. Today's limestone rock was formed on ancient seabeds from the remains of dead marine organisms, mostly algae, corals, and shelled animals. These microbes and animals sequestered calcium carbonate and magnesium carbonate from seawater to form their shells and skeletons. This means that limestone rock is high in calcium and magnesium. This is one of two crucial points that makes limestone water ideal for making whiskey. In the mashing, calcium reacts with grain-derived phosphates and lowers the pH, which stabilizes the amylase enzymes that convert starch to sugar. And both calcium and magnesium are necessary minerals for yeast health. Limestone also effectively filters iron from water. Distillers thin out the strong barrel-proof whiskey (usually 55 to 70 percent alcohol by volume) to bring it down to bottle proof (usually 40 to 50 percent alcohol by volume). If the water they add before bottling is rich in iron, it will introduce off-flavors and turn the whiskey a dark color. Before modern water-treatment technologies, the limestone-filtered water, as a natural source of high-calcium, low-iron water, was ideal for making whiskey. But the advances of water treatment in the twentieth century have made many municipal water sources—not just Kentucky's—perfectly suitable for making whiskey. So, the importance of *limestone-filtered water* (which a geologist would call *hard water*) is much less important now than it was in the nineteenth century.

The second pride of Kentucky bourbon distilleries is their yeast, as almost every distillery cultivates its own proprietary strains. These strains were often isolated decades or generations ago by the distillery's founding members. Right after Prohibition ended, James Beauregard Beam (you may know him better as Jim) caught a yeast strain on his back porch (the distillery had lost their pre-Prohibition strain when the distillery was shut down). They still use this new strain, which they isolated in 1933, today, although it has surely mutated some; cryopreservation was not invented until the 1950s. Each yeast strain is genetically distinct, and just like with different grain varieties, diverging genetics lead to distinct nuances in flavor. During fermentation, no two yeast strains metabolize sugars and other nutrients in identical ways. This leads to a unique production of flavor compounds by each strain. This is why Kentucky bourbon distilleries tout their proprietary strains.

Interestingly, the scotch whisky industry is the exact opposite. Historically, all Scottish distillers have used only the same three to five yeast strains, and they never seriously considered isolating yeasts from the environment or creating proprietary strains. While this is based on guesswork, I believe this is because of the proximity of scotch distilleries to Scottish breweries. Historically, scotch distilleries bought discarded yeast from nearby breweries. (Brewer's yeast is the same species as distiller's yeast, and it's very common for one strain to be efficient at making many different types of alcoholic beverages: beer, wine, whiskey, rum, vodka, and so on.) Once commercial yeast supply companies came into existence in the twentieth century, the scotch industry didn't demand variety, as they didn't believe yeast was anything more than a vessel for converting sugar to alcohol. Occasionally they have introduced and adopted new strains, but only when it shows more efficiency and a higher alcohol yield—it has nothing to do with flavor. (To be fair, over the last five to ten years the scotch industry has begun to rethink its stance on yeast and is now beginning to explore the effects of different strains on flavor.)

The Kentucky bourbon industry, on the other hand, was founded in comparable isolation. If you wanted to make whiskey and there wasn't a nearby brewery, then you had no choice but to isolate yeasts from the wild. Before Prohibition, the title "master distiller" was uncommon, and the head of a whiskey operation was much more likely to carry the title "distiller and yeast-maker."[6] One of my first tasks as a distiller in 2011 was not to make whiskey but to "make" yeast. I spent my first six months on ranches and farms and in labs on a wild yeast hunt. We were successful, and TX Whiskey became the first distillery since Jim Beam in 1933 to isolate and use a wild yeast to make whiskey. Our proprietary strain came from the Rancho Hielo Brazos in Glen Rose, Texas, about an hour southwest of our distillery in Fort Worth. It came from a crushed pecan nut I found lying underneath its mother tree. The strain—which we nicknamed "Brazos"—gives our straight bourbon and rye whiskeys distinct flavors of allspice and fig.

The third thing you'll hear about on a Kentucky distillery tour is the barrel. Federal law requires that American straight whiskey must be aged in new, charred oak barrels. They will most likely tell you about how long they season staves outdoors at the cooperage, how heavy of a char they use, and how different floors in the rickhouse will produce whiskeys with flavors unique to that location.

But when it comes to grain, the details provided on most tours will be relatively limited. Now of course, they will all tell you that by law bourbon must contain at least 51 percent corn. The same simple-majority rule is true with rye whiskey, which must be 51 percent rye. But will they mention anything beyond the very high-level specifications of *yellow dent corn, plump rye, soft red winter wheat,* and *malted barley*?

Before I start to sound like I'm knocking the Kentucky whiskey industry for using commodity, nonlocal grain, let me say once again that there is nothing inherently wrong with making whiskey from commodity grain. Almost every one of the most flavorful and high-quality whiskeys today is made from commodity grain. But the scientific research does support the idea that we can taste the difference in noncommodity, identity-preserved grain. And while I'm not denouncing the overwhelming use of commodity grain in whiskey over the last hundred years, I do believe—based on science and experience—that the move away from commodity and toward terroir will open new doors in whiskey.

* * *

Even though much of the corn used in Kentucky bourbon is grown out of state, there are indeed some long-standing farmer-distiller relationships in Kentucky, and for generations now, some farms have been growing corn and wheat specifically for the bourbon industry. In Loretto, about sixty miles south of Louisville in the heart of central Kentucky, lies Peterson Farms. About forty miles east is Caverndale Farms, near Danville. Langley Farms is just east of Louisville in Shelby County. And Walnut Grove Farm is in south-central Kentucky, only a few miles from the Tennessee border. These multigenerational, family-owned farming operations are good examples of how blurry the line that separates commodity and noncommodity grain can be.

Each of these farms works thousands of acres and grows a variety of crops, often corn, wheat, soybeans, and canola. On the surface, they look like large-scale, commodity grain operations. And in many aspects, they are. What makes them distinct, however, is that much of their corn and wheat harvest is sold to the Kentucky bourbon industry. So is this corn and wheat commodity grain or not?

Sawyer Farms, which supplies TX Whiskey, was originally a commodity grain operation. They still are, although to a lesser extent than before we joined forces. But the grain they sell to TX Whiskey is not a commodity. It has been identity preserved, so we receive only select varieties of corn, wheat, rye, and barley grown by them. Just the same, Peterson, Walnut Grove, Langley, and Caverndale sell some grain to the commodity grain market. And while the corn and wheat they grow is often of the same varieties and grown with the same farming practices as commodity grains, this is the difference: they are not necessarily sold to and stored at a third-party commodity elevator but, instead, stored in silos on their own farms and delivered straight to distilleries. Grain really becomes commodified in community elevators, where it is blended with corn of different varieties and from different farms. The aforementioned farmers may not always segregate varieties, but they do segregate GMO from non-GMO corn, and they do ensure that their grains do not mix with those from any other farms. This is a form of identity preservation. It might not be the most aggressive approach to terroir, but it's undeniable that grains from Peterson, Walnut Grove, Langley, and Caverndale possess provenance, and by extension they can contribute nuances to whiskey that are unique to their environments.

Where the Kentucky bourbon industry has historically fallen short in pursuing terroir is not the fault of the farmers—there are high-quality grains with flavors distinct to specific farms being grown throughout the state. And the established distilleries source from these farms. But they simultaneously source from several Kentucky farms and commodity grain elevators in Indiana. The blending doesn't take place all at once, as the elevators at the distillery can hold only around one week's worth of production. So it's a slow, gradual blend. Multiple truckloads coming in every day, week after week, mingling grain of different varieties and farms. While the grain from Peterson, Caverndale, Langley, and Walnut Grove may capture terroir, it is lost at the distillery.

And then there is rye and malted barley, the usual complements to corn in a bourbon grain bill. Rye is not a common crop in Kentucky, so distilleries source commodity rye from the Upper Midwest, Canada, and even Europe. The malted barley comes from malthouses in the Upper Midwest, as well.

So a meaningful pursuit of terroir has not been pursued by the Kentucky bourbon industry at large since at least Prohibition. That said, exceptions do exist, within both established and new distilleries. Some terroir pursuits are side projects; others are foundations of a distillery's approach to grain selection.

Every year since 2015, Buffalo Trace has planted a different variety of corn on a small farm adjacent to their distillery in Franklin County. Each harvest is mashed, fermented, distilled, and aged separately. After maturation, each harvest will be separately bottled and released as a "Single Estate" expression. While a very interesting side project, these corn harvests represent a very small volume compared to the distillery's yearly output.

Maker's Mark has had a close relationship with Peterson Farms since it began sourcing grain from them after the distillery opened in the 1950s. Peterson is the sole provider of soft red winter wheat to Maker's Mark, and the farm also provides them with much of their corn.

Jeptha Creed, a new distillery near Louisville, has since 2016 used an heirloom corn variety called Bloody Butcher to capture the terroir of Shelby County. Bloody Butcher is a seductive dark red compared to standard yellow dent corn, hence its name. Records of it only go back to the 1840s, when Native Americans introduced it to Virginia settlers, but its origin is thought to go back much further, and it is believed that Native Americans cultivated it for its rich and sweet flavors. Jeptha Creed grows all of the Bloody Butcher corn they use on their family farm. Each year, they keep a portion of the harvest as seed to replant in the next season. I find this approach to be one of the sincerest and authentic methods for capturing and highlighting terroir. This early-generation corn grows right next to their distillery, and by saving seed from season to season, natural and artificial selection will adapt the variety to the very local, specific environment of the distillery's farm. So in essence, I guess you could say terroir plays a role twice: in the grain used to make whiskey and also in the grain that is used as seed.

One of the most impressive new Kentucky distilleries—from both volume capacity and quality standpoints—is Wilderness Trail, in Danville. On a map, you'd notice that Danville is about as central Kentucky as central can be. The town is probably best known as the home of Centre College, a private liberal arts college. In 1921, back when Ivy League schools were known for their American football teams, Centre beat Harvard University 6–0 in what the *New York Times* called "Football's Upset of the Century."

On a Thursday morning in late May, I left my childhood home in Louisville with Leah, my mom, and my uncle Rick (a veteran of the Kentucky bourbon industry) and made the drive southwest to Danville. We were going to visit Wilderness Trail, which I had learned sourced the majority of their grains from only two farms—and that one of these farms was even supplying them with rye. This was the first example of a distillery using Kentucky-grown rye that I had encountered.

15

CORN, WHEAT, AND RYE AMONG THE BLUEGRASS

WILDERNESS TRAIL DISTILLERY was founded in 2013 by Shane Baker and Dr. Pat Heist. They'd been friends since the 1990s and at one time were even bandmates. Their distillery was not some flippant endeavor by two buddies hoping to relive the glory days. At the time of opening, they had already been successful business partners for seven years. Their first company, Ferm-Solutions, is an enzyme and yeast supplier and consultant for the distilling industry. And how did they learn the art and science of distilling? As Pat once told me, "When something goes wrong in the distillery, people always jump to blame the yeast. So Shane and I had to learn every aspect of the process, so that we could explain to a customer why *it wasn't* the yeast."

Shane studied mechanical engineering in college and spent the first part of his career as the VP of engineering for three separate companies. Pat earned a doctorate in plant pathology from the University of Kentucky and was an associate professor of medical microbiology at Pikeville College before cofounding Ferm-Solutions with Shane in 2006. This was right around the time of the most recent Kentucky bourbon boom, which saw domestic sales of the spirit increase 36 percent from 2009 to 2014 and export sales rise 56 percent from 2010 to 2014. The market was hot for bourbon. So in 2013, seven years after opening Ferm-Solutions, Pat and Shane decided to start their own distillery. After only a few short years, their production capacity is now hundreds of barrels a day. They are one of the largest *new* distilleries not just in Kentucky but globally.

Driving east out of Louisville, beyond the outskirts, you enter what Kentuckians call horse country. Scattered throughout the hills are red barns, fences, and practice tracks that constitute the working grounds of Kentucky's other icon—thoroughbred horses. Interspersed throughout the horse farms and hills are rickhouses, which is a type of barrel warehouse composed of long racks where the barrels age and are essentially stacked—or ricked—on top of each other. These are big warehouses, typically holding anywhere between twenty thousand and seventy thousand barrels of whiskey.

We arrived mid-morning, and the first thing I noticed was of one of these barrel rickhouses under active construction. I could already spot four finished ones, and they looked large, undoubtedly capable of storing twenty thousand barrels each. It's remarkable for such a young distillery to have such a large inventory of aging barrels. Most new craft distilleries might produce only a few hundred barrels a year.

My family headed into the visitor's center for a tour while I entered the offices to meet Pat and Shane. Most of their distillery has been built brand new from the ground up, but their offices are in a restored nineteenth-century Victorian-style house that was already on the land. Inside, I found Pat in a meeting room in front of a row of snifter glasses filled with whiskey. Behind them were the sample bottles that laid out their characteristics: style, grain bill, age, barrel, and barn location.

"Just some early morning analysis," Pat said, smiling as we shook hands. Shortly after we caught up, Shane walked in. When you first meet Pat and Shane, you might not immediately recognize them for the highly educated, trained, and knowledgeable expert duo that they are. Their dress is casual, their faces are bearded, and they speak with a warming Kentucky drawl. They make you feel as if you've been good friends for decades, even if you were just recently introduced. But make no mistake—Pat and Shane are two of the most respected figures in whiskey, and they continue to consult for the industry at large.

I'd known Pat and Shane for about five years at the time of this visit. In 2014, after propagating TX Whiskey's proprietary wild yeast strain from freezer stocks, agar plates, and liquid media for every batch, Ferm-Solutions converted our yeast into an active dry form. The unique genetics and flavors of our proprietary yeast strain were the same, but a supply of dry yeast was an incredible time saver. I'd also leaned on them many times over the years for advice when issues arose at the distillery. And while we'd had discussions about grain in the past, they'd never gone much beyond discussing desirable concentrations for starch, protein, and fat.

"So, in our industry, all we've ever heard is, 'You can't grow good rye in the south,'" I said, as we sat there, the row of snifters in front of us suffusing the room with the smell of whiskey. "I've never accepted that, especially since we've had success growing rye in Texas at Sawyer Farms for three seasons now."

When we first started working with Sawyer Farms in 2015, they were already established corn and wheat growers. But after realizing that TX Whiskey had needs for rye and barley, John Sawyer and his team pioneered the growing of those two grains, which was essentially unheard of in Texas. John was able to identify varieties and agronomic techniques that allowed these two grains—typically said to grow better in colder climates—to thrive in Texas. In 2017, Sawyer Farms became the sole supplier of rye to TX Whiskey. Two years later, they became the sole supplier of barley, which is then malted locally in Fort Worth by the craft maltster TexMalt.

"Why is it that rye," I continued, "is grown in such little amounts in Kentucky and that basically all of the rye used to make Kentucky whiskey comes from thousands of miles north? How did you all move away from that system and source 100 percent Kentucky-grown rye?"

"Well, it wasn't always this way," Shane started. "If you look back to the 1800s, there was plenty of rye grown in Kentucky. But when the distilleries came back after Prohibition, there was little Kentucky rye still grown. But commodity rye from up north was easy to get carted in by train."

It turns out that what killed Kentucky rye was what killed the rest of local grain. Before grain was transported across the country by rail, and before modern agriculture shifted farmers away from diverse varieties toward a few high-yielding, subsidized crops, distilleries had no choice but to use local grain. This included local rye for the Kentucky bourbon industry. But during the first half of the twentieth century, Kentucky farmers largely stopped growing rye except occasionally as a cover crop. Today, there are only about a thousand acres of rye grown in the entire state. But enough of those rye acres are at Walnut Grove Farms to meet Wilderness Trail's needs. "The Halcombs at Walnut Grove Farms, which is about 150 miles southwest of us, have supplied us 100 percent of our rye since 2015," Pat said. "It doesn't grow nearly as plump as the rye from up north—the kernels are pretty thin, but they pack flavor."

This issue of size was something I had been considering for a few years already. Like the rye from Walnut Grove Farms, the rye we got from Sawyer Farms was thinner than the northern-grown rye samples to which we compared it. And even for one corn crop that experienced a very dry summer, the kernels were much smaller than normal. Generally, the less optimum the growing conditions, the smaller the grain kernel. The number of kernels on an ear of corn is relatively fixed, usually eight hundred per ear. So farmers usually aim for larger kernels, meaning more yield per ear.

Generally, small, thin kernels also possess lower *test weights* compared to large, fat ones. Test weight is calculated by filling a USDA-approved quart cup with kernels and then weighing it. From this, we determine how many pounds fill a volumetric bushel, which is 1.25 cubic feet. The density of the grain and specifically the compactness of the starchy endosperm can affect the test weight, but so can the shape, size, and packing ability of the kernels. Grain elevators discount grain loads with low test weights, as it is more costly to store and transport low-density grain, given that less weight can be stored in the same volume.

This means that farmers selling to the commodity market want large kernels because they weigh more than smaller ones and have a higher yield per acre. And farmers want high test weights, so they aren't discounted at the grain elevator. But why would distillers care about kernel size and test weight?

Most distillers are trained to mirror the needs of commodity grain elevators, and so they consider test weight a mark of quality. But it isn't, at least not in the context of flavor. It's simply a metric related to efficient storage and transportation of commodity grain. Regardless of how many kernels make up a pound, distillers still pay by the pound. Now, it is true that low test weight can indicate grain with a lower proportion of starch-rich endosperm and a higher proportion of seed coat (the bran). But this is not always the case, and even when it is, that might not be the worst thing, as the bran is full of flavor congeners, such as the hydroxycinnamic acids that serve as precursors to important volatile phenols. When we worked with the lower-test-weight corn from the unusually dry 2018 summer of Hill County, Texas, our alcohol yields held steady. We still produced the same amount of sugar from mashing and the same amount of alcohol from fermentation that we did with high-test-weight corn. The same is true of our thinner rye kernels—test weight is lower than plumper rye from the north, but the starch levels are not significantly different.

As with test weight, kernel size really only concerns farmers selling to the commodity grain market, where yield is the almighty goal. While it might be true that larger kernels look visually more impressive than smaller ones (at least to some people), kernel size is no indicator of flavor quality.

But do small kernels, as Pat put it, pack flavor? After all, in the wine industry, grapes are specifically grown to be small and concentrated in flavor. The large table grapes at the grocery store are nearly half as sweet and double the size of wine grapes. The same is true for many of the crops, poultry, and livestock we eat. Breeding and growing for yield has come at the cost of flavor. Research from Harry Klee's lab at the University of Florida showed that large, modern tomato hybrids contain significantly lower concentrations of sugars, acids, and other flavor compounds than their smaller, heirloom cousins.[1] What *looks* physically appealing to a consumer might not translate to *realized* flavor. Personally, I too found that thin rye kernels (and one harvest of smaller corn kernels) from Sawyer Farms packed flavor, both when they were raw and once we'd distilled them into whiskey.

You'll often hear winemakers talk about how vines must be stressed—for instance, through drought or poor soil—to produce small, concentrated, intensely flavored grapes. There are certain misconceptions with this idea, as too much stress is indeed bad. What *stress* more correctly is trying to convey is that conditions that are *too good* (too much water, soil that's too rich in nutrients in the soil) produce large, healthy grapes that are low in flavor—that is, table grapes. Modern conventional farming has pursued grain growing as it has pursued table grape growing: to bear large fruit. Whether large grain kernels are more diluted in flavor than small ones is an idea that warrants further research. It is true that many heirloom corn varieties have very large kernels and possess great flavor. But might this flavor be even more intense if they are farmed in a way that allows them to "pack" their flavor?

It's a curious question, but it's certainly a concept that every distiller believes applies to oak trees. The white oak trees used to make barrels do not come from ideal soils or environments for fast growth. Instead, they come from places like the Ozarks and the Appalachian Mountains, where the soils are less rich and the trees have to compete for nutrients—they have to *work hard* to grow. Basically, less-than-ideal growing conditions will concentrate the compounds that will become important flavor congeners in seasoned and charred barrels. So, the question for grain then becomes: At what point are ideal conditions for yield—whether those conditions are imparted naturally based on the environment, artificially based on inputs by the farmer, or both—not ideal for flavor?

I asked Pat and Shane this, but they said they weren't sure. "I guess, in theory, there's some validity to the idea," Pat said. "But as far as I know, grain farmers don't consider how certain techniques impact flavor. If they ever did, they haven't for a long time, and especially not for the sake of yield. Or if some still do, they are few and far between."

Here was the dividing line between wine and whiskey that I had encountered almost everywhere since my visit to California. There are the obvious differences between how winemakers and distillers source, select, and assess the quality of grapes and grains. But what I was beginning to realize was that the gulf was wider than that and started even earlier, all the way back in the agronomic techniques used to grow the fruit. I don't mean the decision to farm conventionally or organically, which are more accurately considered holistic approaches to agriculture. I mean how winemakers and grape growers will specifically employ techniques like "vine stressing" that, even at the expense of yield, are meant to improve and concentrate the fruit's flavor. As far as I know, there is simply no equivalent in grain farming. While some farmers—especially the organic and biodynamic ones—are quick to point out that fewer chemical inputs and more interaction with a soil that is alive *could* lead to better flavor, they rarely do something specifically for the benefit of flavor. Instead, flavor is the afterthought, a bonus, considered only once yield and environmental sustainability are assured.

"Stressing the vine" is not a particularly scientific way to describe the techniques growers employ to control and target grape flavor. It also doesn't cover the entirety of those techniques. There's a whole category of agronomic practices in the vineyard—like leaf removal, canopy training systems, foliar fertilization, and bunch thinning—that are intended to influence the flavor congeners as they develop in grapes.

Leaf removal, for example, controls the volume and intensity of sunlight on the grapes. Certain carotenoid-derived flavor compounds like vitispirane (*floral*) and terpineol (*floral, citrus*) will develop in higher concentrations when berries receive more sun. On the other hand, winemakers might not remove leaves or even supply artificial shading to encourage the production of 3-isobutyl-2-methoxypyrazine (*green bell pepper*). While some styles of wines want to avoid methoxypyrazines, certain styles—such as sauvignon blanc, cabernet sauvignon, and those from Bordeaux—owe much of their typical and expected flavors to these compounds.[2]

Canopy training or vine training uses trellising (manmade structures that guide the growth of grapevines) and pruning (cutting of the vine's cordon, which dictates the number of buds—the undeveloped embryonic shoots—that become grape clusters) to control the vine's canopy. The canopy is the part of the vine that is visible above ground: the trunk, cordon, stems, leaves, flowers, and fruits. Canopy training will influence the amount of air and sunlight available to the grapes. This in turn influences the concentrations of terpenes, fusel alcohols, and norisoprenoids in the grape.

Foliar fertilization is fertilizer applied directly to the leaves of a plant instead of to the soil. Sulfur and nitrogen foliar fertilization can change the thiol concentrations in grape berries. For certain styles where thiol concentrations are especially important—such as sauvignon blanc, especially those from New Zealand, which are characterized by the passion fruit and grapefruit flavors that thiols can deliver—such applications are common. Terpenes (*sweet, floral, fruit, resinous*), phenethyl alcohol (*floral, rose*), and phenylacetaldehyde (*honey, rose*) are known to be influenced by whether the fertilizer is proline, phenylalanine, urea, or nitrogen based. And copper, which has been used for pest management since the 1880s, may affect the concentrations of a whole range of flavor compounds.[3]

The amount and timing of irrigation that a vineyard receives has a momentous effect on the development of flavor compounds in grapes. It is true that while drought creates poor grapes, studies show that moderate water stress can concentrate β-damascenone (*cooked apple*), 4MMP (*black currant, box tree*), 4MMPOH (*citrus*), nerol (*floral, fruity*), geraniol (*rose*), linalool (*sweet, floral*), and methoxypyrazines (*green bell pepper, grassiness, earthy*). Certain reports showed that while moderately stressed vines produced berries with higher concentrations of these flavor compounds, it might be more because of the reduction of berry size, another byproduct of water stress. In other words, water stress doesn't lead to more flavor compound production but to small berries, which by nature are more concentrated.[4] However, this might actually support the observation by Pat, Shane, and me that thin rye kernels pack more flavor.

Bunch thinning is the removal of small, overly large, or misshaped grape clusters, and it allows for the most ideal clusters to develop appropriately without crowding one another out or competing for nutrients. Reports show that bunch thinning leads to higher levels of the aromas most desirable in cabernet sauvignon and grenache. One study suggests this is attributable to the increased concentration of the aforementioned 3-isobutyl-2-methoxypyrazine (*green bell pepper*). In sauvignon berries, thinning increased the concentrations of various terpenes (*sweet, floral, fruit, resinous*).[5] While it's the most understudied of the agronomic practices employed to control flavor in the vineyard, bunch thinning is still an important aspect of congener control.

Stressing the vine can improve the flavor of wine grapes, but the situation may be decidedly different in grain. While there is some reason to believe that stressing a grain crop could lead to a higher production of flavor compounds—based on the techniques just discussed with grape growing—it can also cause infections that can lead to the production of mycotoxins, which are harmful to humans and linked to acute poisonings, immune deficiencies, and cancer. Mary-Howell Martens told me about this when we first talked on the phone. "When grain is stressed, it becomes more susceptible to certain fungal infections, and those fungi can produce mycotoxins. So stressed grain plants often do make toxic grain."

Mary-Howell makes a good point about the balance that will have to be achieved if we do explore how *stressing* the plant affects the flavor of its grain. While certain techniques could be explored so that a grain crop isn't basking in an overabundance of nutrients and water, we must do so carefully to avoid inadvertently creating toxic grain.

To wine grape growers, a successful harvest is very much based on techniques rooted in controlling and harnessing flavor. They *grow for flavor*. But for now, when it comes to grains, we do not farm for flavor, at least not like wine grape growers. I asked Pat and Shane how their farmers employ agronomic practices for quality and what exactly quality meant to them and their farmers. "Well, across the industry, quality essentially means that it meets certain requirements for test weight, moisture, high starch, low protein, and acceptable aroma," said Shane. "You don't want any moldy or sour aromas in the grain."

"Sure," I said. The quality marks set by the USDA for the commodity grain trade are the baseline for the American whiskey industry. "That's the same for us at TX Whiskey. We're starting to dive into how different farming techniques affect flavor, but it's still very early on."

I relayed to them that the whiskey terroir research we conducted had revealed clues about chemical markers in raw corn that might correlate with flavor compounds in whiskey. Our results showed specifically that benzaldehyde, carotenoid, and hydroxycinnamic acid concentrations in corn kernels positively correlated with the concentration of many desirable flavor compounds in the whiskey. But we're still exploring what affects benzaldehyde, carotenoid, and hydroxycinnamic acid levels in raw corn, with the thought that maybe one day we can control their concentrations through breeding and farming techniques.

"There is just so much to consider when it comes to flavor, and there are still many gaps in understanding how grain chemistry plays a role," Pat said. "Plus, for decades now, the industry has focused much more on how different oak barrel specifications—seasoning, toasting, charring—and rickhouse environments impact flavor. Barrels have been at the forefront. Grain has always been important obviously, but as long as it looked and smelled good—or just good enough—then that was acceptable."

"We work with two farmers," Shane added. "Walnut Grove Farms supplies us with all of our rye, like I mentioned. Caverndale Farms supplies us with all of our corn and wheat. Caverndale is just down the road from our distillery. Given that they are also a seed producer, they can deliver us extremely clean grain."

Later that day, I would meet Barry Welty, a manager at Caverndale Farms. He told me the same story I had heard from every other farmer I'd met not just since starting this book but since I joined the whiskey industry. The story is one of producing high-quality grains (as set by the USDA) free of foreign material, dust, and off-odors. But it is a story of a farming approach that has been guided by the commodity grain market, not flavor. Again, it doesn't mean that the grain used by Wilderness Trail, New York Distilling, Kings County, and us at TX Whiskey is necessarily commodity grain. But compared to the wine industry, is the grain we use farmed more like table grapes than like wine grapes? I'd say yes.

Changing gears with Pat and Shane, I brought up my chemical roadmap. I also mentioned that while I had discussed it with other distilleries in New York that were focused on terroir, they hadn't been able to weigh in much, as they hadn't conducted any significant flavor chemistry on their whiskeys. I was hoping that Pat and Shane, with their scientific backgrounds and long-term roles as consultants to the whiskey industry, might have some insight. But similar to my experiences in New York, they weren't able to provide much commentary, either. They found the list of flavor compounds interesting, of course, but as far as how to *use* it to better target and control specific flavor compounds, they were unsure. Until we unravel more details about how specific aspects of terroir—grain genetics, farm environment, and agronomic technique—specifically influence the presence and concentration of grain-derived flavor compounds, it will be difficult to make educated decisions on how to harness terroir most effectively. For now, the best we can do is *pursue* terroir and see what flavors we discover in the process.

* * *

Wilderness Trail appears to be one of the largest Kentucky distilleries that can source all their grain from only two farms. One farm not five miles from the distillery—Caverndale Farms—supplies all of the corn and wheat, and the second a few hours to their southwest near the border with Tennessee—Walnut Grove Farms—supplies all of the rye. So they can say definitively that their whiskeys possess provenance, that they express the flavor of a place, and that they are not made from commodity grain. Again—this doesn't necessarily make their whiskeys better than other Kentucky distilleries that rely on commodity grain or that blend many farms and varieties over the course of a distilling season. But it does provide Wilderness Trail with grains that are, to some extent, proprietary in nature, and therefore they will deliver flavors to their whiskeys that are specific to the land from which they came. Flavors from terroir.

* * *

The following day, we didn't have to travel very far to visit the next distillery. We were heading to Rabbit Hole Distillery, which is ten miles from my childhood home, in a newly gentrified part of Louisville known as NuLu. Bordering downtown Louisville to the east, NuLu is a hotbed of trendy restaurants, vintage clothing, craft beer, and—of course—bourbon. If Louisville is Manhattan, NuLu is Brooklyn.

From the inception of Kentucky's whiskey industry, Louisville was home to a number of distilleries. But Prohibition closed the doors on many of them forever. For much of the twentieth century, only Brown-Forman and

Heaven Hill operated distilleries in Louisville. But in the 2010s, with the rise of the craft distillery movement, new operations opened. Near the city center alone, there are now six distilleries along just a two-mile stretch between NuLu and downtown: Rabbit Hole, Angel's Envy, Old Forester, Evan Williams, Michter's, and Peerless. (The downtown Old Forester, Evan Williams, and Michter's distilleries are small satellite operations compared to their headquarter distilleries.)

Rabbit Hole was founded in 2012 by Kaveh Zamanian, an Iranian-born psychologist who grew up in Southern California and spent the first part of his adult life in Chicago. There he met his wife, a native of Louisville. Through frequent visits, Kaveh fell in love with Kentucky. He decided to close his psychology practice and pursue his dream of opening a distillery.

If you go to Rabbit Hole for a tour, their entrance leads directly into the gift store and guest check-in. But immediately beyond the gift store, at the very start of the tour, a flowchart on the wall details Rabbit Hole's whiskey-making process from grain to bottle. This flowchart caught my eye as Rabbit Hole's director of distillery operations, Cameron Talley, led us from the gift store and into the distillery.

"Like many other bourbon distilleries, Rabbit Hole uses the traditional grains of corn, rye, and distiller's malt," Cameron told us. Distiller's malt is the name given to malted barley (usually six-row) that has been produced specifically for the high enzyme content that high-grain mash bills—such as bourbon—require to convert starch to sugar efficiently.

"However, we also use some specialty malts for some of our expressions that are not commonly used to produce bourbon: malted rye, malted wheat, and honey-malted two-row barley."

The *honey* style of malt is produced by the Gambrinus Malting Company in British Columbia, Canada. The origin of the two-row barley for Rabbit Hole's honey malt would certainly be Canada, not Kentucky. I knew that there were no malt houses in Kentucky at the time and that their malted rye, malted wheat, and six-row distiller's malt would most likely also not be from Kentucky. Cameron confirmed this. "All of our malt and rye comes from either Germany or Canada. There are hopes that Kentucky will establish, or rather reestablish, rye and malt industries at some point. But for now, they just don't exist, or at least not in any substantial form. But all of our corn, which is the predominant recipe in our bourbon mash bills, is grown and sourced locally."

I assumed that Cameron meant their corn all comes from Kentucky. But when I asked him to confirm that, he didn't. "Well, no, all of our corn is locally grown. We buy from two elevators, one of which is located just a few miles from our distillery and another about thirty miles east. But some of the corn at the Louisville elevator is grown just across the Ohio River in Indiana. But we're just a few miles from Indiana. This is the Kentuckiana region, after all."

Growing up in Louisville, you hear this word all the time. The local news in Louisville, as the anchors will tell you, broadcasts to Kentuckiana, not just Kentucky. But at first, I didn't understand why Cameron would consider corn grown across state lines in Indiana *local* for a Kentucky-based distillery. But I thought about it more as the tour took us up through the floors of the distillery. Rabbit Hole's tour experience is great for several reasons, not least of which is that you move up their stillhouse—with their twenty-four-inch-wide, fifty-foot-tall copper column still always in view—during the tour, which culminates at their top-floor bar and tasting room.

As we walked up the stairs, I kept turning over in my mind what Cameron said. Up until now, I had been considering *locally grown* to mean that the grain was grown within the same state as the distillery. But the geographical boundaries of states were not, of course, drawn with the concept of terroir in mind. The geographical boundaries of terroir have much more to do with the conservation of climates and soils unique to particular regions.

I considered this idea of regionality in the context of Texas, where I make whiskey. Being such a large state, Texas has incredibly diverse environments—desert, plains, coastal—and, by extension, terroirs. Sawyer Farms is only about fifty miles from our distillery. But let's assume we hadn't found Sawyer Farms and that I was still looking for a supplier. Say I found two farms: one in the Texas Panhandle and the other just across the Red River, in Oklahoma. The Texas Panhandle is hundreds of miles away from the distillery, and the Oklahoma border is about one hundred. Assuming terroir is more authentically captured the closer the farm is to the distillery, then the Oklahoma farm would actually be the better choice than the Panhandle one.

So I asked Rabbit Hole's founder, Kaveh Zamanian. "For us, it really is about *regionality*. Does some of our corn come from Indiana farms? Or, more specifically, from farms in Kentuckiana? Sure. But that doesn't mean it's not local to us at Rabbit Hole. It goes beyond grain for us as well. All of our spent grain, that byproduct of whiskey distillation, goes to farms forty miles east of here in Henry County. And now some of those same farms are starting to grow corn for us. And just a few miles from those farms is where our barrel rickhouses are, aging every drop of whiskey we've ever distilled."

We arrived at the top of Rabbit Hole's distillery and made our way to the window-walled tasting room. I looked out at the Louisville skyline. Just beyond it, underneath the far sky on the northern horizon, was Indiana. We were hovering over Kentuckiana as much as Kentucky. In their pursuit of regionality and of terroir, it made more sense for Rabbit Hole to source some corn from southern Indiana than it does to source from, for instance, Peterson Farms, more than a hundred miles south.

Rabbit Hole's stance on what *locally grown* means made me consider the bigger picture for the corn used to make Kentucky bourbon. A day earlier, I was quick to write off Indiana corn as a travesty against the terroir of Kentucky bourbon. But the situation wasn't so simple. The boundaries of a state are manmade constructs. They don't dictate terroir's boundaries.

* * *

On the third day of our trip, I woke up early and drove east from Louisville on I-64 to Frankfort. People often mistake Louisville, as the largest city in Kentucky, for the state's capital. But it's actually little Frankfort, with its population of 25,000. When Kentucky was incorporated as the fifteenth state in the Union in 1792, a committee of five county commissioners considered proposals from communities who wanted to house the state capital in their town. As the story goes, they chose Andrew Holmes's offer to house the capital in his log cabin for seven years because the proposal came with $3,000 in gold. Frankfort burned down twice in its early years, and both times Louisville and Lexington (the second most populous city in Kentucky) vied to replace it as the capital. But Frankfort, being in the center of Kentucky, retained the honor.

Frankfort is also a whiskey capital of sorts, as it is the home of the Buffalo Trace distillery, well known for making its namesake bourbon as well as Eagle Rare, W. L. Weller, Colonel E. H. Taylor, Blanton's, and the famed Pappy Van Winkle. But when I reached out to them a few weeks before my visit and sent them the chemical roadmap from chapter 11, they demurred. They told me the data they had collected on how terroir affects grain-derived flavors in whiskey were proprietary to their company. Whether this was true or if they even really had any data at all on it, I don't know. I assume they do. Their team of whiskey experts is skilled, extensive, and scientifically minded. Buffalo Trace is well known for exploring the effects of variables in oak barrel maturation, even creating an experimental rickhouse called Warehouse X that allows them to control the temperature, sunlight, and humidity of the aging environment. And they started their "Single Estate" project to explore how different corn varieties influence flavor.

For now, though, my visit to Buffalo Trace was not to peel back the lid on their data but to meet one man, Seth DeBolt, an Australian-born scientist whose career started largely in wine chemistry and plant science. In 2008, DeBolt moved to the horticulture department at the University of Kentucky in Lexington. Familiar with the wine industry, which has developed both extensive collaborations and training programs with many universities worldwide, DeBolt saw that the Kentucky whiskey industry had no such system. In 2015, he founded the undergraduate certificate in distillation, wine, and brewing studies at the University of Kentucky. Its goal was to train students to enter the alcohol beverage industry or at least get them a foot in the door. Of course, with Kentucky's intimate ties to bourbon, many of those students had their sights set on that particular industry. A combination of the program's success and support from Kentucky distilleries has led to the creation of the James B. Beam Institute for Kentucky Spirits (also based at the

University of Kentucky), for which DeBolt was installed as director. According to a university press release, the institute will "educate the next generation of distillers" and conduct research on all aspects of distilled spirits production, including cereal grains.

The Kentucky River runs through the heart of Frankfort, and the Buffalo Trace distillery sits on its banks just north of the city. Off the interstate, the Louisville Road (the old highway that connected Louisville and Frankfort) took me on a hilly path past the state capitol building, over the river, and straight through downtown. Frankfort is quaint; it feels like a city stuck in time. But the people of Frankfort are perfectly fine with that: if they wanted contemporary, they'd move to Louisville.

Exiting downtown, I hugged the Kentucky River's east bank until I arrived at an unassuming brick wall with the words "Buffalo Trace" and the painted head of the buffalo. (At one time, there was an actual trail through Kentucky, Indiana, and Illinois created by migrating buffalo—it was called by many names, one of which was the Buffalo Trace.) The drive through the entrance is shaded by trees and cradled by hills. It claims to be the oldest continuously operating distillery in the United States, and it has gone by many names over the course of its history. While some debate this, its records do indicate that distilling has been happening on its grounds since 1775.

I parked the car and, fighting my natural tendencies, headed away from the distillery buildings and instead up a hill to a playground, where I would meet DeBolt. Soon he appeared over the hill, in a bright blue University of Kentucky polo and jeans, flanked by his two children. With his wavy blonde hair and athletic build, he looked more like an Australian rugby player than an Australian scientist. DeBolt instructed his kids to go play, and if they did so without interrupting, then they would each get a bottle of Freddie's root beer. (This particular root beer is named after the long-time, third-generation Buffalo Trace ambassador, tour guide, and Kentucky Bourbon Hall of Fame member Freddie Johnson.)

I asked Seth how he came to Louisville. "Well, my wife, who is from the States, rides horses, and she would spend summers on the track at Churchill Downs in Louisville. She just fell in love with the area."

Before moving to Lexington and the University of Kentucky, DeBolt earned his PhD in plant biology at the University of Adelaide, in Australia, and his research included collaborative work with the Australian Wine Research Institute. He received his postdoctoral training at the Carnegie Department of Plant Biology at Stanford University. While his research pursuits transitioned from wine to cellulosic ethanol, he maintained academic interests in alcoholic beverages. With the move to the University of Kentucky, the opportunities to reenter alcoholic beverage research emerged.

"The bourbon industry is just really supportive of building academic and research programs that can fill a gap where there has been one for so long now," DeBolt told me. "The James B. Beam Institute for Kentucky Spirits will provide a home for such endeavors. We are working with many of the master distillers as well as the PhD scientists that are on staff at the distilleries."

"So what about you and your lab?" I asked. "Where are your research interests?"

"Well right now, I'm focusing on oak barrel maturation, and working a lot with Harlen Wheatley, the master distiller here at Buffalo Trace. But Chris Morris, the master distiller at Woodford Reserve, is doing some really interesting work with terroir and rye right now."

It turned out that Chris Morris and his team at Woodford Reserve were working with University of Kentucky agronomists and local farmers to bring rye back to Kentucky. The distillery took one variety of rye (the Brasetto variety, which John also grows at Sawyer Farms for TX Whiskey) and grew it on four different farms throughout Kentucky. The plan was to grow, harvest, and distill each farm separately and assess agronomic yield, alcohol yield, and flavor.

* * *

Seth and I discussed Woodford Reserve's project in more detail, but I would also talk by phone a few weeks later with Elizabeth McCall, the assistant master distiller at Woodford Reserve and Chris Morris's protégé. She told me they pursued the local rye project for both flavor and environmental reasons.

"If you consider the carbon footprint of importing rye from farms and elevators thousands of miles away, or even from across the Atlantic Ocean, well—it's just kind of crazy. That was one big reason we decided to pursue this project. Woodford Reserve only works with two Kentucky farms for their corn—Langley and Waverly Farms—so extending the local grain aspect to rye was the next logical step. Plus, farmers used to grow rye in Kentucky, so why not do it again?"

Elizabeth's point was that there is a common misconception that exists for rye and barley—that they grow well only in colder climates. But this is not necessarily true. After all, consider that barley and rye were domesticated in the Fertile Crescent, which is a region in the Middle East spanning modern-day Iraq, Israel, Syria, Lebanon, Egypt, and Jordan. Even thousands of years ago, at the time of domestication, this part of the world was warm.

Elizabeth's next point was exciting to hear, especially given that she worked for one of the largest distilling companies in the world. "Our goal for now is really to explore flavor. Yield doesn't matter, or at least it doesn't matter as much as flavor. Obviously, this will need to work for the farmers too, and we'll need to provide support to them—both from a resource and financial standpoint. But I'm hopeful. The initial results are really promising. Some of this Kentucky rye we've grown seems to have a better, more concentrated flavor than the commercial ryes we buy."

* * *

I left my meeting with DeBolt full of hope and encouragement. Not just specifically for terroir, and not just for the Kentucky bourbon industry, but for the future of whiskey research as a whole. Many well-funded distilleries were pursuing their own internal research, but no company since Seagram's in the early twentieth century had published much of their findings. The James B. Beam Institute for Kentucky Spirits would be based at a university. While there are exceptions, research conducted at a university has a much better chance of being published than research conducted by the industry.

On the drive back to Louisville, I considered the scenery. I had this sense of a deep connection to whiskey all around me. I wondered whether those first Kentucky settlers, whose distilling roots went back to the glens of Scotland and the green hills of Ireland, felt what I felt. Of course, they would have realized that the limestone-filtered water, fertile soil, and extensive river systems that facilitated shipping and transportation were ideal for whiskey making. But was there something else? Something in the air or, more specific to this book, something about the land?

When I lived in Kentucky, before I was a distiller, I was too young to understand what whiskey and particularly bourbon meant to Kentuckians. But I understand it now. Bourbon reverberates through the landscape, the culture, the history, and the people. Kentucky distillers will continue to maintain the integrity and quality of bourbon while at the same time being a driving force in its innovation. If Kentucky distillers decide to commit further to local grain and to the pursuit of terroir, then the positive effects on agriculture and on flavor will be realized far and wide.

Just like in New York, my experiences in Kentucky did not necessarily fill in avenues on my chemical roadmap, but I did find that the pursuit of terroir is alive and well there. Distillers are reintroducing heirloom corn varieties and a forgotten rye crop. Old relationships with farmers are continuing, new ones are being fostered, and a fresh generation of plant scientists are undertaking university-based, industry-backed research.

But perhaps what affected me the most on the visit was something that I struggle to put into words—it's abstract, too much of a feeling. But if I was forced to explain it in one word, it would have to be *reverence*. Kentucky doesn't have a monopoly on making whiskey. Many states have just as ideal ingredients and conditions. But there is a feeling of reverence for the spirit that is—at least for now—unmatched by any other state in the nation. (Although I hope Texas will match Kentucky soon!) Perhaps the pioneering distillers of Kentucky felt that same reverence for what Kentucky whiskey could be. Regardless, it's hard to deny that Kentucky ground feels like more than just a fertile land for whiskey crops—it is sacred.

16

ACROSS THE POND AND THROUGH THE HILLS

When I left Kentucky, my search for terroir had been constrained to corn and rye. In whiskey production, these grains are almost always used raw. Barley is different. It is almost always malted, except in the case of Irish pot still whiskey, which uses both raw and malted barley. Malting transforms raw grain's starchy, hard kernels into ones that are sweet and brittle. It also imparts layers of new flavors to the now malted grain, such as sweet caramel and rich nuttiness. And this raises a question. Does malting negate, mask, or accentuate the flavors of terroir?

I knew I could not totally understand whiskey terroir by considering only raw, unmalted grain. That might be enough for whiskeys produced primarily from grain, such as bourbon and rye. But some of the most famous and commonly drunk whiskeys in the world do not use raw grain. They use malt—specifically malted barley. And nowhere else is malted barley as exalted and championed as in malt whiskey.

The phrase *single malt* evokes a sense of status and quality that surpasses arguably any other whiskey style. (*Single* means only that the whiskey came from one distillery, and *malt* means it was distilled from 100 percent malted barley. Most malt whiskeys are each made at just one distillery, and therefore *single malt* is a more common moniker than malt whiskey.) Malt whiskey is most famously associated with Scotland, Ireland, and Japan, and among those Scottish malts lead the pack.

To clarify, scotch whiskies such as Johnnie Walker, Dewar's, and Famous Grouse are the highest-volume sales of all whiskeys globally and household names to many people. But these are actually blended whiskies, not malt whiskies. Ireland's most well-known whiskeys, like Jameson and Powers, are also blended whiskeys. Blended whiskeys contain a majority (60 percent to 80 percent) of grain whiskeys, but they also consist of 20 percent to 40 percent malt whiskeys (and also pot still whiskeys in Ireland). I've heard blenders use the analogy of painting to describe what they do: grain whiskey is their primer, and they paint flavor on with malt.

I began to search for distilleries that were as focused on terroir in malt whiskeys as those distilleries in New York and Kentucky (and we in Texas) were on terroir in bourbon and rye. I had two destinations in mind—Bruichladdich in Scotland and Waterford in Ireland.

After making whisky for more than 110 years, Bruichladdich closed in 1994. This down period was short-lived, however, and it was revived in 2000 by a group of investors led by Mark Reynier. Reynier was raised in the world of wine, and he was intimately familiar with the concepts of terroir from an early age. He brought this familiarity and pursuit to Bruichladdich, which is one of the most terroir-driven scotch distilleries in operation. Reynier and his partners sold Bruichladdich in 2012 to Rémy Cointreau, and in 2014 he subsequently started Waterford Distillery in Waterford, Ireland. He brought his zealous pursuit of terroir to Waterford, which is not just the most terroir-driven distillery in Ireland but possibly in the world. In a bit of a reverse order, I decided to visit Waterford in Ireland first and leave my trip to Scotland and Bruichladdich for the last leg of my whiskey terroir journey.

* * *

In July 2019, I flew from Dallas–Fort Worth International Airport to Dublin. Leah and I took the eleven p.m. red-eye, which put us in Dublin at one p.m. local time and at nine a.m. in our heads. But I was too excited to sleep. I decided instead to read through the little research that has been published on barley, malt, and terroir. There were three papers—all concerning beer, which is also made from malted barley—and the two oldest were from 2017, which goes to show how new and unexplored this frontier of malt terroir is.

Two of the three papers were authored by Dustin Herb, at the time a doctoral candidate under Patrick Hayes at Oregon State University. Hayes is a plant breeder and professor who leads OSU's barley breeding program, which is nicknamed Barley World. Herb's work explored how barley variety and growing environment influence the flavor of beer. As opposed to my research with corn-based whiskey, where the raw corn itself is mashed, fermented, and distilled, Herb first had to malt the raw barley before mashing.

Malting causes chemical reactions—especially Strecker degradations, caramelization, and Maillard reactions—in the sugars, amino acids, and other precursor compounds during kilning. These reactions will create new and diverse flavors raw grain doesn't have. Now, it is true that these reactions are toned down when the kilning temperatures are relatively low, as is the case for most malt produced for whiskey. The heavily kilned malts that give some beers—such as Guinness—their dark appearance and roasted chocolate flavors are typically not used in whiskey. That said, even low-temperature kilning used for distilling malts will create sweet toffee and caramel notes in both the malt and the whiskey produced from it. Because the flavors of malted barley are mostly believed to develop during the malting process, most research has focused on how different kilning temperatures and lengths change flavor. It has not considered how the chemical composition of the raw barley itself influences those flavors. Instead, distillers and farmers have historically selected malting barley varieties for the characteristics that make them efficient for growing, harvesting, and malting. And just like the other whiskey grains, most maltsters (especially the larger ones) source from grain elevators, which mix together grain of all sorts of varieties and from all kinds of growing environments.

Before the varieties of malting barley make it to the farm and later the distillery, they are screened and approved by associations, institutes, and federal governing bodies. In America we have the American Malting Barley Association. Canada has the Brewing and Malting Barley Research Institute. The United Kingdom has the Institute of Brewing and Distilling. Maltsters are after certain characteristics in barley that will make it ideal for the malting process: rapid maturation, the ability to break dormancy quickly, and uniform germination. They also look for low protein levels as this means more sugar extract. Only certain barley varieties are approved for malting, and growing malting barley requires techniques that growing commodity feed barley does not. Most of these involve maintaining a careful balance of fertilizer to depress protein levels.

But unlike commodity corn, wheat, or rye, governing bodies do evaluate malting barley varieties for flavor. It is, however, one of the last considerations. Those varieties with off-flavors (like sulfur and butter, which result from overly abundant dimethyl sulfide and diacetyl, respectively) are rejected. But for the entire suite of malting barley varieties that are approved, the prevailing mindset says that whatever flavor variations exist among varieties and growing environments are minimal and will eventually be masked by the malting and brewing processes anyway. Therefore, the maltster will swap and blend barley of different varieties and growing environments, assuming that they all meet their quality specifications. Further, the brewer and distiller often don't even know which variety of barley they are buying. Instead, their selection will be based on malt profiles, such as pilsen, pale, Munich, crystal, and Vienna. These different styles of malt can be produced from the same barley. What makes them unique is how the maltster steeps, germinates, and—most importantly—kilns the barley.

There are, however, exceptions. Older, seemingly outdated barley varieties—namely, Golden Promise, Klages, Bere, and Maris Otter—have seen a resurgence. Craft maltsters, brewers, and distillers seek out these varieties because their advocates claim they impart unique and desirable flavors to beer and whiskey that are not present (or not as intense) in modern varieties.

In his reports, Herb cites these varieties and the dichotomy between the wine and beer industries when considering the importance of terroir in general. He hypothesized that something was being missed. The beer industry may well be ignoring a fundamental source of flavor.

The first of Herb's two papers was "Effects of Barley (*Hordeum vulgare* L.) Variety and Growing Environment on Beer Flavor."[1] It explored how barley variety and growing environment change the flavor of beer. "We started this project with a question," he writes. "Are there novel flavors in barley that carry through malting and brewing and into beer?"[2] I was confident that there were, that they were traceable to terroir, and that they would survive mashing, brewing, fermentation, and even distillation. But I wasn't sure about malting.

At the start of his experiments, Herb did something interesting. Instead of working with already existing varieties—as we did in our study—he chose thirty-four varieties from a set of two hundred that had recently been bred by the L. Cistué lab in Aula Dei, Spain, and the lab of his advisor, Dr. Patrick Hayes, at Oregon State. They bred the varieties through a process called *double haploid breeding*—what's important about this approach is that they needed only two parents to generate all two hundred varieties. One of the parent varieties was Golden Promise, which is known for premium flavor. The variety was released in 1968, and even though newer varieties are superior in terms of agronomic performance, Golden Promise persists because brewers and distillers demand it and the unique flavors it harbors. The other parent was Full Pint, a agronomically competitive variety Hayes developed within the barley breeding program at OSU. Full Pint was released commercially in the last decade, and it is gaining popularity with brewers thanks to its unique and desirable flavors compared to commodity malting barley. Developing new grain varieties that have been selected for both flavor *and* yield is, I believe, one of the emerging frontiers of terroir.

Herb grew the thirty-four varieties in duplicate all at one location in Corvallis, Oregon, in 2014. The next year, he grew them across three cities: Corvallis, Lebanon, and Madras. Corvallis and Lebanon are in the Willamette Valley, which receives high rainfall and therefore did not need irrigation. Madras is in central Oregon, which is drier and needed irrigation.

After the harvest, Herb processed all the barley samples at "lab scale," meaning that all the processes of brewing were done in small batches. The technical terms for this are *micromalted* and *nanobrewed.* Partly he did this because there wasn't all that much material to work with, but more importantly lab-scale investigations allow scientists to control variables. Any winemaker, brewer, or distiller will tell you that large-scale operations are predisposed to confounding batch-to-batch factors. In the lab, we can mostly dodge these complications.

Herb did not collect chemical data, but he did gather sensory data. At the Rahr Malting facility in Minnesota, he malted, brewed, and analyzed all the samples. Twelve panelists with comparison-to-reference descriptive analysis quantified the flavors of each sample against a reference beer, the paragon of American lager: Miller High Life.

The panelists quantified the flavors of seventeen sensory descriptors on a scale from zero to eight. If a panelist scored a flavor as a four, that meant the intensity was no different than in Miller High Life. This meant zero to three was "less intense" and five to eight "more intense."

The panelists found significant differences in beer flavor—such as *fruit*, *floral*, *sweet*, *cereal*, *toasted*, *malt*, *toffee*, *grass*, and *honey*—that could only be from the barley variety, its growing environment, or an interaction of the two. This strongly suggested that malting does not mask terroir, at least not in beer. Herb also conducted genetic analyses and identified regions in the barley genome that had significant effects on beer flavor. This suggests that within those regions there are protein-coding genes that play some role in the eventual production of flavor compounds. Unraveling the genetics of flavor in grain may one day aid plant breeders in breeding new, flavorful varieties.

Herb's second paper, "Malt Modification and Its Effects on the Contributions of Barley Genotype to Beer Flavor,"[3] aimed to understand the extent to which malt modification affects flavor in the context of barley variety. In his first study, he used the same malting parameters for every sample. While on the surface this appears to be a sound approach, it is actually a confounding aspect of the experiment. Even though the same malting process was applied to each sample, their individual characteristics would create different degrees of *modification*. In its most basic sense, modification is the extent to which a barley kernel has grown into a seedling. And the degree of malt modification can change the flavor of the beer. So in his second study, Herb wanted to show definitively that barley variety and environment—and *not* modification—were responsible for the significant differences in flavor.

Remember, malting is really just a process whereby the maltster "tricks" barley kernels into thinking it's time to grow into a plant. They do this by creating a suitable environment around the kernel. A kernel of grain is like a prepackaged baby delivery system and incubator. It's composed of the embryo (also called the germ), the endosperm, and the bran. A plant can generate energy via photosynthesis, but a seed kernel cannot. For one, it hasn't yet developed the necessary photosynthetic machinery—not to mention that the seed kernel is in the ground, out of reach of sunlight. But the endosperm is full of energy in the form of sugar, which will feed the baby plant until it shoots above the surface and can photosynthesize for itself.

To pack as much energy into the tight space of the kernel as possible, thousands of sugar molecules are linked together into a crystalline string called starch. While starch is efficient, it is too complex of a food source for the embryo to consume. So when a seed kernel recognizes the conditions are right to begin growth, it begins to germinate. It produces enzymes called *amylases* that chop up the starch chain into simpler sugars for the embryo to metabolize. When maltsters talk about the *degree of modification*, they're describing the extent of enzyme production, starch degradation, and sugar utilization. Part of the maltsters' craft is balancing this so that the malt is *well modified* as opposed to *undermodified* or *overmodified.* Undermodified malt leads to low sugar extraction during mashing and therefore lower-than-target ethanol concentrations after fermentation. Overmodified malt can lead to too much flour during milling, which will cause issues during lautering (the stage in which the mash is separated out into liquid *wort*, leaving the grain solids behind).

Using the same sensory analysis techniques as in his first paper, Herb found that even with the confounding variable of modification, barley variety and growing environment were responsible for meaningful and significant flavor differences.

The third study came from the lab of Adam Heuberger at Colorado State University. Heuberger had many of the same goals as the Herb papers, but he added analytical chemistry to his sensory analysis. Working with six different malt sources, Heuberger and his team investigated the chemical composition of each malt and the beers produced from them. By *malt source* he meant "the combined variation due to genotype and environment, including maltster," and the samples were not micromalted as in Herb's study but obtained from commercial malthouse production lots. Collectively, the six different malt sources were composed of six different varieties (Copeland, Expedition, Full Pint, Meredith, Metcalf, and PolarStar), four growing environments (Montana and Oregon in the United States and Alberta and Saskatchewan in Canada), and four malthouses (Rahr, Malteurop Group, Briess, and Cargill).

Heuberger used a 2.5 hectoliter pilot brew system to brew beer from each malt source. (A hectoliter is 100 liters, and in the paper they fail to mention where all the extra beer went, but I think we can make a safe assumption.) They utilized sophisticated chromatography and mass spectrometry techniques to analyze the flavor compounds of the beers. For sensory analysis, Heuberger gathered a panel from the New Belgium Brewing Company in Fort Collins, Colorado, and used a process called quantitative descriptive analysis (QDA) to assess the presence and intensity of forty-five flavors. While the sensory data generated can look similar to what Herb generated using his sensory technique—in that each flavor is rated on a scale that starts with 0 (not present) and ends at a somewhat arbitrary upper number (8 in Herb's research, 5 for the QDA analysis described here) that denotes the intensity is at its max—the lack of a reference in QDA means that the panelists generate the flavor picture solely from the sample itself. No Miller High Life for reference here. Personally, I prefer using a reference sample; I find it allows me to generate more consistent and meaningful data. But QDA is one of the more proven and established methods for descriptive sensory analysis.

Herb and his panel detected over 246 potential flavor compounds in the six beer samples, belonging to a range of chemical classes: alcohols, aldehydes, alkaloids, alkanes, amines, amino acids, benzenoids, esters, furans, ketones, lipids, organic acids, organosulfurs, phenolics, purines, and terpenes. About 61 percent of the flavor compounds detected showed significant concentration variation depending on the variety, the growing environment, or the interaction of the two. Sensory data showed that compared to the other varieties, Full Pint and Copeland had higher concentrations of *fruity*, *watermelon rind*, *solventy-sweet*, *banana*, and *green apple* flavors; that Meredith and Metcalfe had higher concentrations of *corn chip*, *sulfur*, and *Honeycomb cereal* flavors; and that PolarStar and Expedition had higher concentrations of *umami* and *cardboard* flavors.

Then they correlated the flavor chemistry to the sensory data and hypothesized which flavor compounds might be behind the differences in perceived flavor. For example, Full Pint contained certain nitrogenous compounds and terpenes consistent with the *fruity* and *floral* flavors detected. One such terpene was the norisoprenoid α-ionone (*floral*), which research in 1970 detected in samples of bourbon, rye, and malt whiskeys.[4] Unlike its cousin β-ionone (both of which can derive from carotenoid degradation during high-temperature malting, mashing, and distillation), α-ionone was not detected in *any* of the whiskey sensomics papers I'd investigated. Still, it was another piece of evidence that terpenes such as norisoprenoids will show terroir-mediated concentration variation in whiskey.

Of course, not all of the 246 flavor compounds detected in the Heuberger study will actually influence the flavor of the beer. But the high percentage of those whose concentrations showed significant variation suggests that terroir's influence in beer can survive malting, brewing, and fermentation. And I already had ample evidence that terroir's expression could also survive distillation. So it seemed reasonable that one could taste terroir in malt whiskey. Or, at least, that's what reading the research papers told me. Now I wanted to experience and taste it for myself.

* * *

We landed in Dublin on July 1, and we had not twelve hours earlier been in the brutal heat of a Texas summer. The Lone Star State is a fantastic place to live, but the summers are intense, and the colder-weather respite that this two-week trip would provide was more than welcome.

Wasting no time, our plan was to immediately head to Waterford, in the distillery's namesake city, which is about an hour and a half south of the Dublin airport. Waterford is Ireland's fourth most populous city as well as its oldest—the city is built on a Viking outpost established in the tenth century. A train runs directly from Dublin to Waterford, but the distillery was nice enough to send the company car.

We walked up to the car as Grace O'Reilly stepped out of the driver's seat to greet us. "Hiya," she said. A tall, slender, dark-haired twenty-six-year-old, Grace was born and raised in Ireland, and she is Waterford Distillery's agronomist. The role of agronomist is a rare position for a distillery to have. Her job, as I would come to learn over the next few days, was to be the liaison between the distillery and their more than seventy farmers. She was responsible for monitoring the crops, providing advice where appropriate, and ensuring that the needs and goals of the distillery were realized on the farm. And, of course, she would also bring the concerns, ideas, and goals of the farmers to the distillery.

As we made our way out of the airport, Leah and I got our first taste of left-side-of-the-road driving. From the passenger seat—which to me felt like the driver's seat—I asked Grace a simple question, based on the fact that most distilleries do not hire an agronomist. "How in the world did you end up working for a distillery, and specifically for Waterford?"

"Well, it definitely wasn't some goal that I set out to achieve," Grace said, laughing a bit. "I got a degree in agriculture at the University College Dublin, and I got to know Mark Reynier over the years before he offered me the job at Waterford."

"Why did you go into agriculture?" I asked.

"I grew up on a farm about one hour north of Dublin. Both my dad and grandfather were farmers. Up in the northern part of Ireland we don't grow as many crops. We raise cattle and sheep, mainly."

Grace majored in animal and crop production as an undergraduate. She spent time across all sorts of farming enterprises, including a sheep research center in Ireland and a dairy farm in New Zealand, and conducted crop performance trials. "It was during my time spent in the fields working on field trials that I became interested in growing crops. My lecturer spotted this interest and mentored me to pursue a career in crops."

Grace eventually landed a job with Minch Malt. In operation since 1847, Minch is Ireland's oldest and largest malthouse. It's located about sixty miles north of Waterford in County Kildare. (There are thirty-two counties in Ireland; Waterford is located in County Waterford.) Minch provides barley seed that they've independently certified to more than six hundred screened and approved malting barley farmers across eight different counties in Ireland. So many farmers and all the logistical and technical situations they pose in ensuring each field and each crop has the best chance of achieving the malting barley quality specs means that they need full-time agronomists to work with the farmers all year round, whether they are planning, planting, cultivating, harvesting, storing, or transporting. Grace was one of these agronomists.

"With Minch Malt, it was my job to *walk to crops*," Grace told me, "meaning I would go into a field and assess the quality of the barley. It was my job to do everything I could to make sure the farmers would deliver malting-quality barley."

The quality specifications for malting barley basically come down to ensuring the crop has a sufficiently high germination rate, is free of disease, and is acceptably low in protein. The percentage of protein is critical, as high protein levels in barley will produce low alcohol yields in the distillery. This is because higher protein levels mean lower starch levels, and starch is what is converted into sugar, which is in turn fermented into alcohol. Less alcohol in the beer translates to lower alcohol yields in distillation. Given that all-malt recipes like single-malt whiskey rely solely on the malted barley for potential sugar, it's imperative that protein levels stay sufficiently low. (Side note: For recipes with a preponderance of grain—like bourbon and rye whiskey—it's actually desirable to malt barley that is high in protein, as this will translate to more amylase enzymes in the malt, which we need so that only a small percentage of malted barley in the grain bill is sufficient for converting all of the starch in the grains into sugar during the mash.)

"Acceptable protein levels range from 9.5 percent to 12.5 percent," Grace told me. "But Waterford was always seeking protein levels on the low end of the spectrum."

It turned out that Grace worked *with* Waterford Distillery before she worked *for* them. While she was at Minch Malt, many of the farmers whose crops she walked grew for Waterford. She even appeared in promotional videos for them while she was a Minch Malt employee. She explained to me that the distillery's process is unique in that every year they execute forty to fifty "single-farm" distillations.

Single-farm distillation—this was a new term for me. *Single barrel* was well established in our industry—it meant all the whiskey in the bottle came from the same barrel. *Single farm*, I knew, was analogous to *single vineyard* in wine. But I asked Grace to clarify the *distillation* part of the phrase *single-farm distillation*.

"Every week, we distill malt that was produced from the barley of a single farm. In all, we do about forty single-farm distillations every year. We usually have contracts with around seventy farmers, but we know that not every farm will produce malting-quality barley, and some might fail to have a harvest at all. The gist is, though, that we separately buy and store barley from around forty different farmers every year. Then throughout the year, we individually send barley from each farm to Minch, who turns the barley into malt, and then trucks it down to our distillery for distillation."

There it was—single-farm distillation. The polar opposite of commodity grain distillation.

Grace told me that in 2018 she moved from Minch to Glanbia, a multinational nutrition group with a yearly revenue of over 3.5 billion euros. There she focused on wheat grown for feed, selling fertilizers and pesticides to the farmers whose crops she walked.

"But Mark was persistent," Grace said. "He got back in touch and asked me to join Waterford. I decided to take a chance on his offer, atypical as it was for an agronomist. I officially joined this past February." I asked Grace why she decided to leave Glanbia and a more traditional role for

Waterford. What changed and made her decide to take Reynier up on his offer the second time? "When I joined the Glanbia team, I was exposed to many different types of crops throughout different regions of Ireland. I really came to grasp what terroir meant and what it could potentially mean for whiskey. As Mark and I talked, we both agreed it was now or never for me, and I took a leap of faith!"

As we got farther away from the city, we connected with the M9 motorway, a seventy-four-mile equivalent of an American interstate highway that runs just south of Dublin all the way to Waterford. The Irish countryside in all of its green glory overtook the landscape. Rolling hills were not just a cliché—they were everywhere.

"We work with some of the farms around here," Grace said. "The bulk of our farmers are located between Dublin and Waterford. The soil here, as well as ample sunshine, is good for growing barley."

I looked out and saw fields as far as I could see. Some held grazing cattle and sheep, while others had rows of crops. I noticed that some of the crop fields were decidedly more green than the others.

"See that field there?" Grace asked, as she pointed to one of the browner fields. "That's winter barley." Barley is split into different categories at specifications that are more generic than the variety classification. One such category is based on the rows of seeds on the barley head, also called the spike or the ear. There are two-row and six-row varieties. Another category is based on the growing seasons, winter and spring. Wheat and rye have the same growing season classifications for its varieties. In Ireland, winter barley is planted in October and harvested in July, whereas spring barley is planted in March and harvested in September. The exact time of planting and harvesting can vary based on the weather, but usually only by a few weeks. Grace pointed out that the reason the winter barley fields were browner than the spring barley fields was just that they were closer to harvest.

"Spring barley is the main type used for malting in Ireland," Grace said. "It's one of the easier crops to grow and can achieve some of the highest yields in Europe. It thrives on our temperate climate with mild, moist summers and lack of temperature extremes like frosts or droughts. For the maltster, under the right climate conditions, spring barley meets the quality specifications easier than a winter-sown barley."

The time passed quickly with the conversation and views of the Irish countryside out the window, and before long, we arrived at the outskirts of Waterford. We took a small city-center bridge across the River Suir. From it, we had a prominent view of Waterford Distillery, nestled up against the river. Its façade was long and shaped in a way that almost made it seem as if the building itself was sailing down the river, like the egg-shaped Gherkin building in London, except lying on its side.

Grace dropped us off at our hotel, a mere two hundred meters from the distillery doors. Leah and I found a nearby restaurant, where we drank fresh Guinness (it does taste better in Ireland) and ate our first of many meals of fish and chips with mushy peas. After tipping what is normal in America but very abnormal in Ireland—or so I gathered from the waitress's face—we headed back to the hotel. Before 10 p.m., on a day that started in northern Texas and ended in southern Ireland, we were fast asleep.

17

TĒIREOIR

THE NEXT MORNING, after enjoying our first traditional Irish breakfast, we made the short walk to the entrance of the Waterford Distillery. The buildings are painted in shades of white, steel gray, and baby blue. I would later find that this scheme echoed that of Reynier's first distillery, Bruichladdich. I never did ask if that was coincidence or contrivance.

At the entrance gate, we buzzed the security guard, who appeared to be expecting us. He opened the magnetic gate and led us to the distillery grounds. He introduced himself as Eamonn, invited us to take a seat in the lobby, and told us that Grace and Waterford's head brewer, Neil Conway, would be down shortly. In the lobby, a large painting covers a wall's entirety with the word "TĒIREOIR" printed in capital letters.

Before long, Grace and Neil emerged from a door and greeted us with a friendly warmth that would become very familiar—the Irish are incredibly welcoming, and they make you feel at home even if you are thousands of miles from it. Neil was a few inches shorter than Grace, with a freckled face and black-rimmed glasses. Like me, he seemed like the type of brewer who would be equally comfortable in the lab as in the brewery or distillery. Just as with Grace, I asked Neil how he came to work at the distillery.

"Well, I actually started working here in 1998, when this whole facility was a Guinness brewery," Neil told me. For most of its history, the distillery grounds had housed a succession of breweries, Guinness being the most famous. Later that morning, Eamonn the security guard would transition into a fantastic tour guide (a good example of how employees at small distilleries must wear many hats), and he took us around the abandoned brewery, which ceased operations in 2003, when Guinness built the new egg-shaped brewery.

The first brewery on the grounds was established in 1792 by Davis, Strangman & Company. It changed hands many times over the years, but much of the existing equipment was over a hundred years old, and some of the wooden ceiling beams and windows dated back to the original eighteenth-century construction. Perhaps the most exciting things we saw were the nineteenth-century open-top wooden fermenters with interior copper lining. These are called coolship fermenters, and the copper lining allows for better thermal conductivity. Before modern heat exchangers were introduced, this characteristic was important for cooling the beer between mashing and fermentation. Leah is obsessed with copper, so she was hooked on them right away. And so was I. Open-top fermenters are familiar to me; they're all I've ever used to make whiskey. But they're a rarity in breweries these days, aside for notable exceptions like those used to produce Belgian lambics.

In 1948, Davis, Strangman & Company closed their brewery, and Guinness purchased the site in 1955. Neil became a brewer there in 1998 and stayed until 2010, when Guinness closed the plant to consolidate operations at their headquarters at St. James Gate in Dublin. Neil followed the brewery to Dublin and brewed Guinness for another five years until he decided to leave the industry. It would be a short-lived departure. In 2015, he moved to Glanbia, the same company that Grace worked for after Minch Malt and before Waterford.

"At Glanbia, I was a shift manager at one of their new dairy nutritional ingredients plants. But in 2017, a few years after Mark had bought the brewery and started Waterford Distillery, Ned called me and offered me a position as head brewer." Ned Gahan was the head distiller for Waterford, and I would meet him shortly. Like Neil, he had spent a career making Guinness in the same building where he now made whiskey, at least once Mark

Reynier bought it. "When Bruichladdich sold to Rémy Cointreau in 2012, Mark decided to leave the company," Neil told me. "He wanted a fresh start, a clean slate—he wanted to build his vision from scratch."

Mark Reynier is what we might call a maverick in the whiskey industry. I knew he had revived Bruichladdich in 2000, but besides that I wasn't too familiar with his story until a few months before our visit. In February 2019, *Forbes* interviewed us both, as well as Bruichladdich's Simon Coughlin, for an article on terroir in whiskey. Coughlin is Reynier's former business partner and now the CEO of Global Whisky Brands at Rémy Cointreau. I mentioned the book in the interview, and shortly after Waterford's head of communications, Mark Newton, reached out to me to see if their approach to terroir might be worth investigating further. Mark extended an offer to visit their distillery in Ireland, and he also gave me Reynier's contact information. So I phoned him.

A frank, tireless speaker, my one-hour call with Reynier consisted of him talking for about fifty-five minutes of it. But I enjoyed every second. His excitement, tenacity, and vision for Waterford—and for the pursuit of terroir at large—was infectious, and I felt it again among all the distillery's employees.

Before venturing into whiskey, Reynier had been a third-generation wine importer in London. "Growing up, for Sunday lunch, my father made us guess the wine being served, and the justifications for our guesses were always rooted in the idea of terroir." Reynier told me how he had transitioned from the wine to the whiskey industry and summarized the history of agriculture in postwar Europe. What he told me confirmed my suspicions about how high-input, chemically intensive agriculture masks the effects of terroir. "After World War II, the introduction and heightened use of agrochemicals caused some wines—even those from Burgundy—to lose nuances from terroir. When you grow grapes for yield and feed the vine chemicals, the vine doesn't grab the bedrock." So even grape growers sometimes pursue yield at the expense of flavor.

Reynier is as zealous about terroir as anyone I've ever met. In 2016, he even began a new rum operation—the Renegade Rum Distillery—which is as equally committed to capturing and highlighting terroir in rum as

Waterford is for whiskey. It was Reynier's passion—and the opportunity to return to his old brewing home—that convinced Neil to join Waterford. "When Ned called me up to pitch the job, I was at first not entirely sure," Neil said to me in the distillery's welcome center. "But after talking to Mark, I quickly accepted the position."

Neil joined Waterford Distillery in 2017, about two years before my visit. As head brewer, he focused largely on those parts of the process from incoming malt to the distillation of new-make. Ned, as head distiller, was equally in charge of distillation, but he also spent much of his time monitoring and blending maturing stock. At many distilleries, Neil would carry the title of distiller and Ned the title of blender. But titles—and the actual roles conducted by any given position—are flexible from distillery to distillery.

Before we went inside, I asked Neil about the spelling of terroir painted on the wall. "Ah, yes," Neil said, grinning and looking toward the mural. "That was Mark's idea. It's a play on words with *eire*, which is Irish Gaelic for Ireland. But it's become more of a word that we use to neatly encapsulate our production philosophy—like a stamp of our approval, to demonstrate the integrity of our barley. So we now aim to use TĒIREOIR more explicitly in the labeling of our whiskey, so that it serves a practical purpose."

We followed Neil from the welcome center and into the distillery, stopping right inside the entrance. On the wall were five individually framed flow diagrams laying out Waterford's process with imagery and in detail: barley, malting, mashing and fermentation, distillation, and maturation. Neil walked us through the process.

Every season, Waterford contracts with forty to seventy farmers—each with a different soil type (there are around twenty distinct types across all the farms) and farming approach (conventional, organic, and biodynamic)—to grow barley just for them. Typically, about ten different varieties are grown across all their contracted farms. They don't choose these varieties for flavor necessarily; instead, they're usually dictated by the Irish Department of Agriculture, and it's based on available seed and suitability to certain soils and regions. But Waterford is also conducting variety trials with Minch Malt, focusing on heirloom varieties. For the 2019 season, they are growing Hunter and Gold Thorpe. The former is showing promise, but the latter grew too high, causing the stalks to fall over, a condition called *lodging*. Aside from the difficulty lodged stalks create for harvesting, when the plant tips over and the ear nears the ground, the kernels can take up moisture. This moisture can trick the kernels into thinking they've been planted, and they can prematurely germinate on the ear.

Every farm aims to harvest one hundred tons of barley. Each farm's harvest is sent to a local grain handling facility, where it is dried and stored in a building Waterford has coined the Cathedral of Barley. (An interesting side note: not every climate requires commercial drying machines—some regions, like Texas, are hot and dry enough that the kernels dry out on the ears before harvest.) Inside their cathedral are forty separate bins, and each stores barley from a single farm. Throughout the year, Minch will malt each farm's barley separately before sending it to Waterford to store in their silos. Every week, Waterford then separately mills, mashes, ferments, distills, and barrels each farm's barley. They use a double-pot distillation system, which is somewhat unusual given that Irish whiskey commonly uses triple distillation.

One of the pot stills they use today was formerly on display in front of Bruichladdich when Reynier worked there. When Rémy Cointreau bought them out, Reynier cut a deal to take the still with him. He had it refurbished and recommissioned at Waterford. Their whiskey comes off the still and is barreled at around 140 proof (70 percent alcohol by volume). They barrel every farm's barley separately—about 200 to 220 barrels per farm—and they use a range of different barrel types.

For every farm, 50 percent of the cooperage is used bourbon barrels (for many Scottish and Irish distilleries, 90 percent or more of what is produced is aged in ex-bourbon barrels), 20 percent is new (which they call *virgin*) toasted or charred oak barrels, 15 percent is ex-wine French oak barrels, and the remaining 15 percent is a mixture of different used sweet—often fortified—wine barrels, for example, port, sherry, madeira, and sauternes.

Neil walked us through the distillery to show us the process firsthand. On their main platform, where the pot stills, mash cooker, and mill are all in close proximity, there is a large window that looks out over the River

Suir. The bridge we crossed when we first came into Waterford the day before was easy to spot. Now we were inside the belly of the egg.

Neil explained Waterford's single-farm distillation program, which is truly one of a kind. Yes, they're tapping into the flavor of a region by forgoing commodity grain and working directly with farmers, but they're taking the whole concept one step further. By capturing and highlighting the flavors of individual farms, their new-make sees inherent and desirable flavor changes from week to week, given that the barley each week originates from a different farm. This is decidedly different than how, for example, we capture and highlight the flavors of only Sawyer Farms at TX Whiskey. The new-make we produce sees inherent, small drifts in flavor from harvest to harvest, but throughout the year, in between the harvests, the flavor of the new-make stays consistent. Of course, Texas is large: 130 million acres of farmland compared to Ireland's 11 million. There are farms large enough in Texas to supply most if not all of the grain needs of even the larger distilleries. In Ireland (and Scotland), this is typically not the case. Most of Waterford's farmers grow thirty-five to forty acres of malting barley yearly. Some distillers might have reservations about the idea of purposefully introducing flavor variation from week to week, but Waterford sees it differently. "We're chasing that variation," Neil told me. "Week to week, we strive to keep every part of the process consistent, but that is so the flavor nuances we yield are truly from the barley itself. Those nuances, you see, well, that's terroir. That's what we are after."

Terroir is more than just an aspect of Waterford's approach to whiskey making—it's the foundation on which everything else is built. That's why they felt it was important to do what they could to prove that it is real—not only that it exists in whiskey but that its existence is profound. To further this mission, they teamed up with an impressive list of scientists to form the Whisky Terroir Project.

* * *

If you visit Waterford Distillery, they'll almost certainly allow you to taste samples of their new-make whiskey produced from barleys of different terroirs, each highlighting how variety, farm, and agronomic technique can influence flavor. According to them, flavor differences among terroirs are "detectable even to nondrinkers." I agree with them. Some of their new-make samples are full of bready, cereal notes, while others have prominent fruity and floral notes. But even so, Waterford conducted a scientific study to prove it, for the same reasons we did.

We had two goals for our bourbon terroir research. First, we wanted to determine for our own sake if modern corn varieties—even with their relatively narrow genetic diversity—are still affected by terroir and whether these effects translate to flavor chemistry. If the answer was no, then we wouldn't need to be so strict about the varieties we used or the farms from which we sourced them. And if we did find that terroir changed flavor, we wanted to publish our findings in a peer-reviewed scientific journal. This would be a way to fight for the integrity of the idea of terroir, which we believed was more than just a marketing buzzword.

This second goal was also what motivated Waterford's study. Reynier was fully confident—from his time in the wine industry and then at Bruichladdich—that terroir mattered in whiskey. But given that some of the larger alcohol corporations had dismissed and even belittled the idea, he decided to produce definitive proof.

Before 2019, I didn't know about Waterford, and Waterford didn't know about me or TX Whiskey, but we were both conducting research with nearly identical methods and nearly identical goals at nearly the exact same time. Our paper on terroir in bourbon was published in 2019, and Waterford's own scientific paper on terroir in barley whiskeys is set to be published in 2021. We started our projects around the same time, but Waterford decided to incorporate the variable of the "season," what winemakers would call the vintage. Our study looked only at corn grown and harvested in 2016; Waterford analyzed three different years. Because of this, they are still in the midst of the project, analyzing barley from the 2017 and 2018 seasons. But they have finished their analyses for the 2016 crop.

They say the whiskey industry is incestuous. For instance, the Beam family has been involved with nearly every long-standing distillery in Kentucky, not just the Jim Beam distillery. This seems to be true for agricultural scientists too. At the start of my trip, I read some of the most up-to-date research on whiskey and terroir, most of which was written by Dustin Herb, of Oregon State University's Barley Breeding Program, under the advising of Patrick Hayes. Before his doctorate, Herb received his master's degree in plant breeding at Texas A&M, and my own doctoral advisor, Seth Murray, sat on his thesis committee.

When Herb published his two papers on terroir in beer—effectively showing that the barley variety and growing environment did change the flavor of beer—Reynier realized that the same techniques could be used to "prove" that terroir in whiskey was also real. He reached out to Herb to see if he had an interest in joining their research team. The lab work was to be conducted in Ireland and Scotland, but Herb was a staff researcher and plant breeder at OreGro, a seed production company in Oregon founded by his family. He couldn't physically be at the lab, but he agreed to join the team as lead data analyst and author. He was the author for the Whisky Terroir Project's preliminary results, which were published ahead of the forthcoming journal article.

The project was a joint venture that brought together teams from Ireland, Scotland, and the United States. Waterford began the project in 2016 by growing two different barley varieties (Olympus and Laureate) in replication at two different farm sites—Athy in County Kildare and Bunclody in County Wexford—a little more than thirty miles apart. After harvest, they analyzed the barley for sixteen different micronutrients. While the variety did not show statistically significant differences, the farm location did. Barley from Athy had higher levels of selenium, chromium, and tin, while barley from Bunclody had a greater concentration of barium, cadmium, zinc, copper, and aluminum. What these micronutrient differences might mean for flavor we don't know, but it was an interesting find, and something my own study hadn't explored.

After the four barley samples (Olympus–Athy, Olympus–Bunclody, Laureate–Athy, and Laureate–Bunclody) had been stored for a few months to break seed dormancy, they sent them to Minch Malt for lab-scale malting. There they worked to produce a standardized base malt to limit the effect of the malting process. Even though the malting process was standard, genetic differences between the varieties or as a result of the growing environment meant barley from Athy was more modified than Bunclody. Chemically, this meant the Athy barley had less beta-glucan, less starch/sugar extract, and more protein.

The malted barley samples then took to the air, shipped by plane to the famed whiskey consultancy lab Tatlock & Thomson in Scotland. There, the lab's director Dr. Harry Riffkin oversaw the lab-scale mashing, fermentation, and distillation. His team took the new-make whiskeys and conducted quantitative descriptive analysis, which was the same sensory analysis technique—described in the previous chapter—employed by the Colorado State University professor Adam Heuberger for his beer terroir research paper. They found that new-make whiskey produced from barley grown at Athy had more *fruit* and *malt* flavors compared to Bunclody. Conversely, new-make from Bunclody barley had more *sulfur* and *grassy* flavors. They did not find significant differences between the two barley varieties, but they did find significant effects from the interaction of variety and growing location. The Laureate barley grown at Athy had big flavors of *cereal* and *dried fruit* flavors that the other three samples didn't. Once Tatlock & Thomson had finished their analyses, they sent the new-make samples back to Ireland for analytical chemistry.

The great roadblock to this science, as we saw in the United States, is funding. It's expensive to conduct the sort of hard-core research that can prove the influence of terroir. But Waterford was able to receive funding for the chemical analysis from Enterprise Ireland, an Irish state economic development agency that supports innovative Irish-owned businesses to accelerate their development and position in the global market. Their funding made it possible to bring in a postdoctoral researcher for the analytical chemistry, which would be conducted at Teagasc under the guidance of Dr. Kieran Kilcawley. Teagasc (which is the Irish Gaelic word for "knowledge") is the government research and development, training, and advisory service in Ireland for the agri-food sector. Kilcawley is a flavor chemist who focuses mainly on dairy, so the Whisky Terroir Project interested him not

TABLE 17.1 RESULTS FROM WATERFORD DISTILLERY'S WHISKY TERROIR PROJECT

CLASSIFICATION	FLAVOR COMPOUND WITH SIGNIFICANT DIFFERENCES IN WATERFORD'S NEW-MAKE ATTRIBUTABLE TO TERROIR[1]	IMPORTANT FLAVOR COMPOUND IN BOURBON, RYE, AND/OR MALT WHISKEY[2]	AROMA	ORIGIN(S)	INFLUENCED BY TERROIR IN NEW-MAKE BOURBON[3]	INFLUENCED BY TERROIR IN WINE
Aldehyde	Acetaldehyde	Bourbon/rye whiskey	Green apple	Fermentation, maturation	Partially	Wine[4]
Ester	Ethyl laurate	Malt whiskey	Floral, waxy	Fermentation	Yes	Wine[5]
	Ethyl lactate	Not reported	Butterscotch		Not identified	Wine[6]
Furan	Furfural	Not reported	Almonds, baked bread	Grain, maturation	Not identified	Wine[7]
Fusel alcohol	n-propanol	Bourbon*/rye whiskey*	Musty, solvent	Fermentation	Not identified	Wine[8]
	n-butanol	Bourbon†/rye whiskey†/ malt whiskey†	Balsamic		Yes†	Wine[9]

*Important as a potential flavor compound precursor.

†Technically not n-butanol but its isomer isobutanol.

Sources can be found in appendix 4: "Key to the Roadmap: Sources for Chapter 17."

just for its implications in whiskey but for how terroir might affect feed barley destined for cattle—and therefore the flavor of dairy products. But the terroir of milk is definitely outside the scope of this book!

Kilcawley and Waterford recruited Dr. Maria Kyraleou, who for her PhD research studied how different viticultural techniques change the phenolic composition of red grapes. The team will analyze the flavor chemistry of the new-make whiskey samples using gas chromatography and mass spectrometry. This was exciting to me, as it meant that maybe—after little luck in New York and Kentucky—someone else might finally be able to add to my whiskey terroir roadmap. Would they report that terroir influenced many of the same flavor compounds that we had detected in our research or (via our homology comparison to beer and wine) hypothesized to be influenced?

While Waterford will publish the bulk of their findings in 2021, they have shared some preliminary results with the public through their website. This is what they found. To follow along, table 17.1 summarizes their results and indicates how they overlap with the whiskey sensomics and bourbon terroir research from previous chapters.

Considering the effects of the environment, they found that concentrations of acetaldehyde (*green apple*), n-propanol (*musty solvent*), n-butanol (*balsamic*), and furfural (*almonds, baked bread*) were significantly influenced by the farm at which the barley was grown. Concentrations of furfural, for example, were significantly higher in new-make whiskeys distilled from barley grown at Athy compared to the same barley grown elsewhere. The same for n-propanol in barley from Bunclody. They also detected significant differences for variety: new-make whiskeys distilled from the Olympus variety had a higher concentration of ethyl laurate (*floral*). For the interaction of the environment and the variety, the team found that ethyl lactate (*butterscotch*) concentrations showed significant differences among the new-make whiskeys.

TX Whiskey and Texas A&M found something similar for acetaldehyde, whose concentration was at least partially affected by the farm environment. Further, it was negatively correlated with the total "good" aroma units.

We found similar results for ethyl laurate, an ester. They found that the variety caused significant differences in concentration; we found it was the interaction of the variety and the farm. Further, it was reported that ethyl laurate is an important flavor compound in scotch malt whisky.[1] Finally, while we did not detect ethyl lactate, and neither did any of the whiskey sensomics papers, the Waterford and Teagasc teams did—and I'm not surprised. Ethyl lactate is the ethyl ester of ethanol (produced by yeast) and lactic acid (produced by bacteria). Studies have shown that the makeup of the indigenous lactic acid bacteria on malt can depend on where the malt comes from and that these lactic acid bacteria produce different flavor compounds (including lactic acid) during fermentation.[2] The fact that the Waterford team detected significant differences here from the variety-farm interaction has a hypothetical explanation.

While we didn't detect n-propanol in our study, some of the whiskey sensomics research[3] detailed earlier in this book found that the ester ethyl propanoate is a potentially important flavor compound in—at a minimum—bourbon and rye whiskey.[4] Ethyl propanoate is the ethyl ester of ethanol and propanoic acid, and propanol can oxidize to propanoic acid under the right conditions. So, while it's a stretch to connect n-propanol and ethyl propanoate, it is at least worth considering. Likewise, we did not detect n-butanol in our study, but the whiskey sensomics work showed that its closely related isomer isobutanol (isomer meaning it has the same chemical formula but a different arrangement of its atoms in three-dimensional space, if you remember back to the penny, nickel, and quarter analogy from chapter 10) is a potentially important flavor compound in bourbon, rye whiskey, and malt whiskey. Likewise once again, we did not detect furfural in our study. But we did find that over 37 percent of the variation in the concentration of 2-pentylfuran (*fruity, grassy*) among the new-make bourbon samples was attributable to the farm environment. Both furfural and 2-pentylfuran belong to the class of flavor compounds known as furans, and while it takes a few steps of chemical reactions, furfural can serve as a precursor to 2-pentylfuran.

Perhaps one of the most important items to highlight, though, is that for all of the flavor compounds detected by Waterford to be affected by terroir, there are multiple reports in the academic research literature showing that terroir affects their presence and concentration in wine (table 17.1). Wine continues to serve as a model system for which to study terroir in whiskey.

Waterford drew the same conclusions I did in my study. The grain variety, the growing environment, and the combined interaction of the two had a statistically significant effect on many important flavor compounds. Or, more concisely, two separate whiskey-making teams were independently finding that terroir does indeed affect the flavor of whiskey.

18

CULTIVATING FLAVOR ON THE FARMS OF ÉIRE

AFTER SPENDING MUCH of the previous day with Neil discussing how their distillery operated and covering the status of their terroir project, I was eager to get out into the countryside and visit with some of Waterford's farmers. Along with tapping into the soil, climate, and topographical variations that exist throughout the farms they work with, Waterford's cadre of growers includes farmers pursuing conventional, organic, and biodynamic techniques. While I had experience with the different approaches to farming—largely through my interactions with those farmers in Texas, New York, and Kentucky we've met previously—Waterford offered a unique situation that might help me better understand directly how these factors affect flavor. Did Waterford's growers farm in a manner so that flavor in the grain was paramount? Or were they still primarily focused on those quality specifications established for commodity malting barley? Even though the farmers I'd previously talked to sought to produce the highest-quality grain possible—oftentimes with identity preservation—their methodologies weren't always aimed at how to heighten flavor in grain, much less select for particular flavors. Now, in defense of these farmers—and all other grain growers—the current scientific landscape has not effectively elucidated how different farming techniques influence flavor in grain, much less the whiskey that is made from it. Still, I was curious to learn more about how Waterford's growers approached barley farming, knowing that their buyer was obsessed by both flavor and terroir.

After another Irish breakfast at our hotel, Leah and I made the short walk back to Waterford, where we were met outside by Grace and the company car. We were heading about an hour north to a town called Durrow, in County Laois, to the farm of Seamus Duggan, one of Waterford's more successful and tenured growers. Duggan farmed conventionally, Grace told me, but that didn't mean he was using chemicals excessively or that he didn't strive to "feed" his farm through nature. "Seamus raises cattle too," Grace said on the drive through more rolling green hills into the countryside. "His fields are rotated, incorporating grains, beets, and grass. Beets help maintain soil structure and increase soil aeration, and the cattle graze the grass. He also uses manure from the cattle to fertilize his fields, so that his use of synthetic fertilizers is minimized."

It was a picturesque day to visit a farm—sunny and just warm enough you didn't need a jacket but not so warm you'd sweat. "Sunny Ireland" is mostly an oxymoron, but not so in the southern part of the island. They call it the "Sunny South East." The locals certainly enjoy the sunshine (realizing of course that it is still Ireland and that rain is expected every day or two), and so does the barley. Sunshine promotes the *grain fill*, which is the term for when the kernels themselves begin to grow.

We pulled up to Seamus's home at the entrance to a set of his fields. As we got out of the car, Seamus appeared from around the back of his house. A tall man with a sturdy build and sporting jeans, work boots, and a baseball cap, Seamus looked like many of the farmers I had met in the United States. Both his shirt and hat were green, and he almost blended into the scenery of his farm. After brief introductions, we made our way down a paved path to one of his barley fields, which was no more than a hundred yards from his home.

At the entrance to the field was a gate with a sign that read "Waterford Distillery: Barley for Single-Farm Distillation." I would see these signs on nearly every one of the farms growing barley for Waterford, usually on a public-facing gate or wall. It was the same sense of pride I saw in the farmers from Texas, New York, and Kentucky, the pride of knowing their barley would be distilled—individually—into a whiskey that could be tasted and shared with friends and family. The signs weren't just promotional pieces for Waterford—they were badges of honor for the farmers.

"When Waterford first approached me with this idea," Seamus told me, "Well . . . I thought it was far-fetched, as did many of the other farmers from that first year or two." We were out in the fields talking, Grace inspecting the crops a few feet away. From his matter-of-fact tone—and Grace's nonreaction—I could tell he wasn't revealing some deep secret. The Waterford team didn't just have the skepticism from certain drinks corporations to contend with. Even the local farmers had their doubts at first.

"Barley was barley, or that's what I used to think." Seamus said. We walked through the field, running our hands through the barley heads. The stalks were about waist high, and the heads glistened and swayed in the sun and wind. "But Waterford made me see it differently. I mean, I guess I always understood the general notion of terroir. Even within my own farm, the unique characteristics of each field require slightly different farming approaches. But the differences in flavor—that was a surprise."

Seamus told me that it wasn't until he tasted new-make malt whiskey made from his barley and his neighbor's barley—not more than fifteen miles away—that the reality of terroir made a lasting impression in him. "The flavors were totally different," he said. "With that taste, well, that's when it hit me—there's a grain of truth to this terroir thing." In my notes, I circled the "grain of truth" quote from Seamus and complimented him on his play on words. Grace said he must have had it planned all along.

Seamus also explained that Waterford paid their farmers more per pound for barley than all other potential outlets—assuming, of course, that it met Waterford's quality specifications. With no other high-paying outlets competing for their barley, it wasn't exactly like Waterford's farmers had a seat at the table, like some winegrowers do. But it's a step in the right direction. Waterford's farmers are paid for quality—and paid well.

As I inspected Seamus's barley more closely, I could see that the heads were already well formed and heavy. If it wasn't for the strong, short stems of this modern barley variety, the plants would have probably lodged at this point. Breeders selected for stronger and shorter stems specifically to hold the weight of fatter heads. The older varieties often didn't have this characteristic, and so are prone to lodging. Seamus told me he'd had firsthand experience with this when he was responsible for handling the heirloom varieties Waterford was trialing at a nearby farm. The barley field was on a hill that overlooked much of his entire farming operation. Many of the other fields were dotted by his three hundred cattle. Seamus was a conventional farmer, so he used some synthetic inputs—fertilizers and pesticides—to grow crops and raise cattle. His fields were, at least to my eye, completely free of weeds. Knowing this is a sensitive subject, I delicately asked him about his approach to farming and how he viewed the use of synthetic inputs in conventional farming versus the natural ones organic and biodynamic farmers use.

"Well, it's not like we just spray and treat without any care or thought," Seamus said, a bit defensively. "Our goal is to produce a good crop and to do so without exhausting our land. I think there is a misconception that conventional farmers care only about the former and not the latter. But the reality is that many of us are just as concerned about the innate fertility of our soil as anyone else. What good would my land be from season to season if I didn't care about its health?"

As we walked back to his house, I asked him whether he employed any techniques that fell under the category of "farming for flavor." "Well," he said, "we are growing a variety called Laureate, which is said to have a great flavor. Our main goal from a farming standpoint, though, is to encourage low protein, which we do by applying just the right amount of nitrogen at the right time. If the protein is too high, then Waterford won't accept it, and we then have to sell it to another distillery or brewery with less strict requirements or potentially even as feed. And of course, we do everything we can to avoid disease, like fusarium. But I don't know of any specific techniques for actually targeting specific flavors. If there were any, I'd surely try them out."

Laureate is a modern, two-row spring variety that was bred and selected by Syngenta, a Swiss agricultural biotechnology company. The variety has gained prominence in the brewing and distilling industries, although like all modern varieties, it wasn't bred with flavor as a primary selection criterion. But it is one of the highest-yielding malting barley varieties that is free of glycosidic nitrile. This is important, as glycosidic nitrile can serve as the basal precursor compound for a potential carcinogen called ethyl carbamate. Most whiskey distillers will monitor their spirit to ensure that the levels of ethyl carbamate are sufficiently low. But the problem can be avoided altogether—at least in malt whiskey—by the use of barley varieties free of glycosidic nitrile.

Back at Seamus's house, he brewed a pot of tea and brought out a plate of scones and a side of fresh butter. (The Guinness isn't the only thing that is better in Ireland; the butter is equally and distinctly delicious.) Then he brought out a bottle of whiskey. It was young, only about a year old, but it still had the lovely amber color of oak maturation. It was whiskey made from Seamus's barley.

"Would you like to try a glass?"

It was the look on Seamus's face when he offered me the whiskey that convinced me that growing barley for Waterford was about much more than just the premium they paid. "Sure," I replied.

Seamus's face lit up, and he promptly poured us two glasses. "This here is the best whiskey made so far, right, Grace?"

"Aye, Seamus—it's the best," Grace said, laughing and smiling.

The whiskey was good, especially given it was only a year old. It had a clean, malty flavor base, but layered on top were the distinct fruity notes from some of the French wine casks that were used in maturation. But its flavor alone wasn't what captured the moment for me. Here I was, in the middle of rural Ireland, drinking tea, eating scones with fresh butter, and drinking whiskey made from barley grown not more than a hundred yards from where I was sitting. And I was enjoying it with the farmer himself, basking in his pride. It was the expression of a very specific place. Seamus himself said it best: "We are the primary producers of a grain that ultimately makes a whiskey or a beer. But never can you say that your name is actually attributed to one particular bottle. Except for here, my name on this bottle, that came from my crop and my field, which is what it's all about—provenance. To me, that's fantastic."[1]

The very real provenance of that glass, the connection between Seamus and his land, that's what made the tasting experience go beyond just flavor. Whatever it transcended *to* is hard to put into words, but that itself might be a true mark of terroir. There are indeed tangible, scientifically grounded aspects to terroir, but there's also something ineffable, maybe even mystical, and the importance of that aspect was becoming more and more clear to me. Maybe every aspect of terroir couldn't be scientifically explained. But I was starting to realize that was all right . . . because you know it when you taste it.

* * *

After departing Seamus's farm, we headed farther north, to County Tipperary. Landlocked and rural, Tipperary is a farming town nestled among hills, rivers, and lakes. We were heading there to visit Waterford's most well-known farming duo—Ross and Amy Jackson, the Jacksons, as Grace collectively called them. Ross is Irish and sports a mop of curly red hair, and Amy is English, with straight, blonde hair. Their distinct accents and appearance are an attractive contrast that I decided must have helped make them into the public-facing farmers for Waterford that they are. What's more, Ross is a twin, and his brother Alan farms next door to him. He also grows barley for Waterford.

As we drove up, Grace told me that the Jacksons' farm started as a conventional farm but eventually turned into one where every crop and animal is raised organically. "Amy could explain more as to why they made the switch," Grace said, as we drove up their gravel driveway and spotted Amy off in the distance corralling sheep.

We walked over to Amy as she regimented the sheep, and then after quick introductions we headed toward their barley fields. On the walk, I asked her why the farm had switched from conventional to organic. "Well, one reason is definitely the premiums that organic demands," she said. "And Ross's brother Alan had already made the switch to organic, and it was working out good for him. Plus, we could just tell that the flavor was better with organic. We don't eat our organic barley—the price is too good!

We sell that to Waterford. But we do eat some of our organically raised meat, and compared to conventional, the flavor really is better. We also realized that our conventional way of farming had consequences. We wanted to work with the biology of our fields, not against it, and keep the soil as healthy as possible."

I'd seen a promotional video from Waterford where Ross said, "It's interesting, you can actually tell the difference between good soil and bad soil by the smell of it."[2] That smell is real. When the soil is alive and teeming with microbes and organic matter, it smells earthy and slightly sweet. When the soil is compacted, it won't drain well, the oxygen becomes depleted, matter decomposition slows, and it develops a sour, ammonia-like odor. Tilling is the common way to break up compacted soil for planting, but tilling can decrease the concentration of microbes and organic matter by exposing them to air, which facilitates mineralization and erosion. To avoid compaction, farmers won't use heavy farm machinery when the land is wet. If they do till, or even just when they want healthy, living soil, they'll often rotate crops or grow cover crops to refresh the soil with organic matter. When they made the switch from conventional to organic, the Jacksons grew grasses and red clover to restore fertility to their soils.

As we were walking around their farm, even stopping to enjoy some raspberries growing in their garden, an SUV drove up. It appeared almost as suddenly as we could hear it. It pulled up next to us, and out hopped Ross. Like every farmer I'd met, he was wearing jeans. But Ross was also the youngest farmer I'd seen, and it appeared that younger meant more casual. He'd ditched the collared shirt for a T-shirt. After another set of hellos, we walked to the field where Waterford's barley grew.

"We rotate our fields to ensure that the soil stays healthy," Ross said, "and it looks like this: We'll grow barley on a field back to back, and then we'll do oats, then grass, and then vetch. That system keeps the soil fertile." Vetch is a legume in the pea family, and therefore it can fix nitrogen from the air during growth. Grasses have an extensive root system that pulls nutrients from deep down in the soil. Rather than harvesting them, the Jacksons plow the grasses and vetch back into the ground, replenishing the topsoil with nitrogen and nutrients from the air and the deeper soil.

The Jacksons were unsure how exactly healthy soil improved crop flavor. But their hypothesis—which I believe is sound—is that nutrients in the soil are building blocks for the plant's metabolism. If the soil is low in nutrients, there is less for the plant to build from and therefore less diverse and concentrated byproducts (such as flavor compounds) of the plant's metabolism.

Ross must have seen my eyes drifting toward the weeds in the barley field, which were more noticeable than those at Seamus's. "Yes, we do get more weeds than conventional growers. But we keep them at bay, especially with cover cropping. Ultimately, some weeds grow here, but they're not a big deal." This was something I had considered previously. Many farmers—especially conventional ones—have been trained to kill every last weed on their fields. In some respects, this is good practice. Weeds compete with the crop for nutrients, and an overabundance will depress your yield. But there must be *some* point where the presence of *some* weeds isn't a serious concern. A few weeds scattered here and there throughout his crop were apparently not something that worried Ross.

"You know," I said to Ross, "I've been wondering if perhaps some competition in the field might actually lead to the production of more or different flavor compounds in a crop." The idea goes something like this. Plants talk to one another, but not with words. They talk through chemicals, like in the 2008 movie *The Happening*, in which (spoiler alert) plants began to communicate with one another through airborne compounds, and those compounds included an airborne neurotoxin that drove humans to commit mass suicide. This isn't so far-fetched. Plants do communicate through a range of different volatile organic compounds (VOCs). It may be to communicate the location of a nutrient source or an approaching predator, but whatever the prompt, the plant will release VOCs into the air and through the soil to notify neighboring plants. Sometimes the VOCs are toxic to the predator—like insects or other animals—and double as a defense mechanism.

Take this one step further and consider flavor. If a crop is growing on a field with no weeds, no bugs, and plenty of nutrients (whether natural or synthetic), then it wouldn't need to communicate much with its fieldmates.

But if some weeds came around—not enough to compete but enough to make their presence known to the crop—then that might encourage the crop to produce VOCs. These VOCs might serve as flavor compounds or trigger the production of flavor compounds. It's just a hypothesis of mine, and not one I plan on researching anytime soon. But it's one potential explanation for why organic crops taste different. It might not just be the soil but also the chemical communication that exists in a field with more biodiversity. After all, monoculture was not the environment in which our crops were first selected and domesticated. Ross and Amy found the idea intriguing, but they wouldn't necessarily be encouraging the growth of unwanted plant species anytime soon—weeds were still weeds.

During our visit with the Jacksons, we didn't try the whiskey distilled from their barley. But Ross has visited Waterford to smell and taste the whiskeys in the past. And was his barley the best? Being humble, he wouldn't answer me directly. But when Waterford asked him on that same promotional video if he really thought the whiskey from his barley was the best, his response was, "Well, look, it's quite subjective, but that's my opinion." Ross said this while laughing, but you can tell that the answer also came from a place of pride, confidence, and friendly competition. And this may be one of the most important aspects of what Waterford is doing. Their farmers are essentially competing with one another. Friendly competition, sure, and as long as they all get paid fairly, nobody takes it too seriously. But it's also a real artifact of Waterford's single-farm-distillation approach.

And why is this important? Because just as competition among plants can potentially lead to new flavor compounds, so can competition among farmers. This competition will drive innovation, and it will drive change. Whether it's new varieties or new farming techniques, Waterford's farmers will explore every angle to get that edge, and all for bragging rights, so that they can say their barley makes the best whiskey. So, sure, on the surface, it's just a fun competition. But I'm convinced it will drive new farming innovations in an industry that is otherwise slow to do so. For now, I wouldn't say Waterford's growers are exactly farming *for* flavor, but they are farming with flavor in mind.

19

AT LAST, A SIP

We spent our last day in Waterford with their head distiller, Ned Gahan. Like Neil the head brewer, Ned had also worked at the Waterford distillery back when it was a Guinness brewery. We met Ned in his sensory lab. Tucked away downstairs in a relatively dim area of the distillery, from afar Ned's domain looked a bit like a mad scientist's basement lab. But as we got closer, we saw that the lab's wall of windows revealed a brightly lit white room lined with shelves of sample bottles filled with amber liquids.

Ned greeted us, smiling: "Are you ready to do a proper tasting?"

Ned has been at the helm of Waterford since the beginning. When I asked him how he had become Waterford's head distiller, he told me that when he first met Mark, he had no idea who he was or what he'd done with Bruichladdich. "I met Mark first in 2014 when he was visiting the old brewery site. I knew he was looking to buy it and turn it into a distillery. So I came in, showed them around the brewery, and spoke with them about how it had operated. A few months later, I contacted him about any potential job opportunities. I went to meet him, we had a chat, and he offered me the job. It was my responsibility to help turn the old brewery into a state-of-the-art distillery."

Ned was working on a blend when we walked into the lab. Waterford will release limited extensions that highlight single farms, and potentially other bottles that highlight agronomic techniques or barrel types, but the ultimate goal is to fashion their main brands in the same manner as cuvée wine blends. Ned is responsible for deciding how to combine their whiskey stock—consisting of spirits from more than forty single-farm distillations every year, matured in four different general barrel types, and of different ages—to create, as Reynier puts it, "the most complex whiskey possible."[1] Waterford would ultimately release their first whiskeys to the market in 2020.

Behind the counter where Ned was blending up a new trial blend were rows of glasses filled with new-make whiskey. Behind them were their bottles, each labeled with the farmer's name, county, soil type, farming technique (whether conventional, organic, biodynamic), and barley variety. This was the first time since our TX Whiskey/Texas A&M research that I could compare the influence of terroir across a range of whiskeys. I had never tasted whiskeys produced at the same distillery—using the same mashing, fermentation, yeast strain, and distillation methods—where the only variables were tied to grain variety, farm (or even farming region), or agronomic technique. And even with my own research, all of the corn had been farmed using conventional methods—this would be my first time ever tasting the variable of conventional versus organic versus biodynamic.

I started with three samples made from conventionally grown barley. Two of them tasted very similar, which made Ned nod a bit as if that was to be expected. They weren't overly complex, but they had crisp fruity flavors and subtle cereal notes. "Those two samples were made from barley grown in the same county, in very similar soil types and climates," Ned told me.

The third sample made from conventional barley was distinctly more flavorful than the first two. Where the first two had subtle cereal notes, malt and toffee flavors were prominent in this third one. Next I compared this third, conventionally grown sample to samples made from organically and biodynamically grown barley. I was surprised at how equally intense the toffee aromas were in all three. The Waterford team had told me that whiskeys from organic and biodynamic barley usually packed more punch, but my many discussions with conventional farmers told me that just because they might use synthetic inputs doesn't mean they overuse them or ruin the delicate balance of microbiotic life in the soil. In some cases, to achieve balanced, healthy soil, conventional growers will use many of the same techniques that their organic counterparts do.

While the third, conventional new-make whiskey had toffee notes similar to the organic and biodynamic ones, it lacked some of the complexity. The organic had notes of tropical fruit where the biodynamic had floral notes of roses. While these flavor differences could have been attributable to the different agronomic techniques employed, all three were produced using different barley varieties and in different counties. So, technically, I couldn't tell whether the minor differences in flavor were from agronomic technique or something else.

The sixth sample I tasted was made from an heirloom variety of barley called Hunter. It was grown by none other than the inimitable Seamus Duggan. This one stood out from all the rest. Not necessarily because it was better. In fact, it was less fruity and delicate than the others. But it did possess the most intense flavors of nuts and toasted cereal of the group. Waterford has no plans to distill whiskey from this variety again; it was an experiment. But they'll keep exploring heirloom varieties, expecting each to harbor unique flavors worth highlighting. Next up in their pipeline was a variety called Goldthorpe, introduced at the beginning of the twentieth century and said to possess great flavor.

* * *

Now, at this moment, you may be asking yourself: So when is the "eureka moment"? When will he reveal the secret formula: that Variety X grown in Environment Y using Agronomic Technique Z makes the best whiskey in the damn world? How is that chemical roadmap he keeps talking about going to reveal the secrets of terroir?

If so, I'm sorry, but that eureka moment never came. I don't know whether I was even expecting one. (Maybe a little, especially at first.) But as my journey continued, I realized that I'd never find such a straightforward answer. By the time I was tasting at Waterford, I had already discovered that terroir wasn't some phenomenon that changed an easily circumscribed number of flavor compounds. There is a multitude of potential flavor compounds in whiskey that come from the chemical constituents of the grain. Some of them come directly, like certain terpenes and aldehydes. Others come indirectly and act as fermentation precursors, as with esters and fusel alcohols. Essentially every single chemical constituent in grain can be influenced by the genetics of the variety, the environment of the farm, and the agronomic approach of the farmer. Terroir isn't just a part of grain development and composition, like one section of an orchestra. It's the conductor.

Earlier in this book, I alluded to an important point—the current scientific literature, at least in my interpretation, does not suggest that terroir in whiskey will lead to vastly *better* whiskeys. But it can lead to unique nuances in the flavor of whiskeys. It can concentrate flavor, especially when it means using new or forgotten varieties bred and selected for flavor and growing these varieties in the particular environments that allow their flavors to be most intensely and truly expressed. Certainly that would give us a more diverse and concentrated set of flavor compounds than what we find in commodity grain. And terroir can potentially unlock and reveal flavors that have been forgotten (and maybe never experienced) in grain and whiskey. Imagine the entire suite of flavors possible in whiskey as rooms in a mansion. Terroir is the key to some of those rooms—maybe even whole corridors—many of which have been closed up for a very long time. But what we cannot expect from terroir is a map to making the *best* whiskey. For one, as Ross Jackson said, *best* is subjective. Second, terroir is about capturing and expressing the flavor of a specific place, and there is no *best* place. Provenance is not the same as preference.

But before me were laid out samples of all sorts: whiskey made from organic barley versus conventional, from heirloom varieties and modern ones, from farms in the southern part of the island and farms just outside

Dublin. The differences between them, those flavor nuances that existed, was terroir.

And those distilleries that are actively pursuing, capturing, and highlighting terroir are rewarded with those flavor nuances and the innate distinctiveness that accompanies it. Varieties may be shared among distilleries, but growing environments may differ. When growing environments are the same or similar, varieties can be different. And even when they are not, every field within a single farm has its own character. Terroir is to a distiller what a proprietary spice box is to a chef.

Perhaps I wasn't that much closer to identifying exactly which flavor compounds were responsible for the flavor variations I had experienced in Waterford's whiskeys or for understanding how the varieties, soil types, climates, topographies, and agronomic techniques controlled the presence and concentration of those compounds. I had some sound hypotheses, but I still didn't know how the forces of terroir controlled the presence and concentration of flavor compounds in whiskey. In many ways, I wasn't much closer to understanding terroir than I was when in the California wine country. Except for one thing.

The science wasn't far enough along to explain how terroir worked, and it may never be, at least not in one cohesive theory. Simply put, terroir is extremely complex. Not to mention it's subject to individual interpretations. (My interpretation, backed by science, is laid out in this book, but it is still only my own.) But what the science does support, and what I had discovered through my journey, was that terroir in whiskey was real. Terroir *does* change the flavor of whiskey, and the flavor diversity of whiskey will grow as distillers pursue, capture, and highlight terroir.

Will this change the overwhelming use of high-yield, indistinguishable, commodity grain to make whiskey? Personally, I hope so, for the sake of flavor, and the farmers, and the environment, however small a role we distillers have in that. But that's a lofty, probably pie-in-the-sky hope. Commodity grain will still dominate the global grain trade even if every distillery switched to locally sourced, identity-preserved grain. And as long as commodity grain is around, which I suspect it will be for the foreseeable future, much of the whiskey on the market will be made from it.

But if we, the whiskey industry and its consumers, accept and appreciate terroir—and in doing so recruit distillers, maltsters, farmers, and plant breeders to champion the cause—then we can change the future of whiskey. Whiskey could enjoy a diversity of flavor, a meaningful existence of provenance, and a connection to the land that we have lost over the last hundred years.

20

THE CHURCH OF SCOTCH WHISKY

After nearly a week in Ireland with the Waterford team, Leah and I made our way to the Dublin airport. But we weren't flying back to the United States, at least not yet. We had one more stop to make—Scotland, and the Bruichladdich distillery, located on the Rinns, the western peninsula of Islay.

A rugged and mountainous island with an area of only 240 square miles, Islay is the southernmost island of the Hebrides, an archipelago off the west coast of mainland Scotland. Getting to Islay is no easy task. One way is to take the ferry from Kennacraig, a hamlet on the Kintyre peninsula, about a two-and-a-half-hour drive from the Glasgow airport. The other way is to fly direct from Glasgow or Edinburgh. We flew from Dublin to Glasgow, where we boarded a small Twin Otter turboprop to Islay.

From the airplane window we looked out at the peninsulas and small islands that dot the ocean between Glasgow and Islay. Scattered throughout these are peatlands, which are essentially waterlogged areas with conditions (low pH and low oxygen) that preserve vegetation and organic matter in a state of partial decomposition. Because of its high carbon content, they call peat "the forgotten fossil fuel." When it's dry, it barely smells. But when burned, it gives off its distinctive smoky, medicinal, and phenolic aroma. In whiskey, we call the aroma from peat just that—"peaty." The flavor compounds from peat are introduced to the barley during the malting process. Many of the Islay distilleries use barley that is kilned by burning peat instead of natural gas, and the smoke imparts that characteristic flavor into the malt. This is why Islay is known for producing peaty single-malt whiskies.

Laphroaig, Lagavulin, and Ardbeg are all situated on the same road just miles apart, and they produce some of the more famous, heavily peated Islay scotch whiskies. There are six other distilleries on Islay, and all but one of them are known for producing peated whiskies. The one outlier is Bruichladdich.

In truth, Bruichladdich does produce one of the peatiest whiskies in the world (a brand called Octomore), but they are perhaps better known for their name-brand whisky, which is produced without peated malt. But that's not why I was visiting them. Whether they use peat or not, the distillery's approach to whisky and flavor had a single foundation upon which everything else was built—terroir. This was no surprise, since Waterford's terroir-obsessed Mark Reynier was also behind Bruichladdich.

This was our first time in Scotland. And while I was excited, I was a little nervous. Not about being in a new country, or being away from home for another week, or flying in a tiny tin-can turboprop across an expanse of cold sea and densely forested mountains. But instead because of a question that was haunting me. As I looked out the plane window, I kept wondering: What was left to learn from Bruichladdich about terroir?

This question didn't come from a place of cockiness. Every time I visit a distillery, talk with its employees, and observe its operations, I learn something. I knew my visit to Bruichladdich would be no different. But I had already visited many of the most terroir-driven distilleries in the world and read all the research on it that had ever been published. I also knew Bruichladdich's approach to terroir was more, shall we say, philosophical than scientific. This isn't meant as a slight—Bruichladdich was our forerunner, the first modern distillery to really pursue terroir in the making of whisky. But unless Bruichladdich was undergoing research with academic collaborators that I didn't know about, then it was unlikely they would have anything concrete to add to my chemical roadmap. The chance to taste their whiskies that highlighted different varieties, growing environments, and agronomic techniques would be meaningful, but I already had enough evidence to say definitively that terroir influences the flavor of whiskey.

So I could guess what I'd find. Their whiskies made from 100 percent Islay barley would taste different from those made from mainland Scottish barley. Their organic barley whiskies would be distinct from those made using conventional crops. Their whiskies made from heirloom Bere barley would taste different from those made from modern, high-yielding varieties. Maybe not every flavor difference would be attributable to terroir. Some might come from innate batch-to-batch variations during milling, mashing, and distillation. Some might come from the different barrel types used or the number of years the whisky was aged. But *some* of the flavor would be attributable to terroir. Every time a thread is tugged, the whole tapestry shifts.

I was looking forward to our adventure in Islay, our engagements with Bruichladdich, and tasting their whiskies. But I wasn't confident that they'd give me anything more I could use in this book.

This was a short-lived concern. My experiences at Bruichladdich, while maybe not rooted in hard science, would reveal something else about terroir—maybe the most important thing about terroir. Not a scientific revelation, but something rooted in the connection between people, the land that they nurture, and the whisky they share.

* * *

We landed mid-afternoon at the Islay airport, which was nothing more than a strip of runway alongside a terminal the size of a diner. After being handed the keys to our rental car by a laid-back Scot who seemed far less worried about my left-side driving than I was, we loaded up our small hatchback and set out. I never became fully comfortable driving during our stay in Islay, but after a few minutes on the road, mirror-image driving instincts do begin to emerge. *Just keep the grass on your left side*, I would tell myself.

Our destination was not a hotel exactly but one of the houses that Bruichladdich owns near its distillery. Ours was called An Taigh Osda (a Scottish Gaelic word I tried to pronounce once—and then never again). It was even closer to the distillery entrance than our hotel in Waterford had been.

The Bruichladdich distillery takes its name from the small village in which it resides on the Rinns. Islay is split into two parts. On the eastern side, spanning the length of the island's most northern and southern parts, is the "mainland" (which is almost amusing, since the island runs only twenty-five miles north to south). The western part of the island, which is shorter than the mainland, is the Rinns peninsula, or simply the Rinns. The airport is on the western side of the mainland, and Bruichladdich is on the eastern side of the Rinns. The roads in Islay are all either narrow two-lane roads or even narrower one-lane ones, on which one car has to yield to the other.

Hesitantly and bravely (at least in my own mind), I drove us north on the A846, stopping when we hit the administrative capital of Islay, Bowmore. Sitting at almost the same latitude as Bruichladdich—which is just a few miles west as the crow flies across the tiny Loch Indaal sea inlet and visible on a clear day—the town of Bowmore is most famous for housing the Bowmore distillery. Established in 1779, it is one of the oldest distilleries in Scotland. Even though its population is under a thousand, the town of Bowmore feels like a burgeoning city, as most of the other towns, villages, and places of interest on Islay are—in a welcoming way—extremely quiet. A town of one thousand on Islay feels like downtown Manhattan.

We continued around the northern part of Loch Indaal, entering the Rinns peninsula, and then south to Bruichladdich. The terrain of Islay is lush hilly forests, green pastures, and coastal dunes. It's intoxicating and mesmerizing. It stirred in me the same feelings I felt when I visited Kentucky. Being on Islay was like stepping into the church of scotch whisky. Its atmosphere and the smells, hotels and restaurants, hills and seas—the omnipresence of whisky permeated the island.

We pulled into An Taigh Osda's driveway around dinnertime. Sharon McHarrie, Bruichladdich's director of hospitality, was there to greet us. She gave us a brief tour of the house and led us to our room on the third floor. This was the top floor, and out the window we could see Loch Indaal. A long and wide beach met the seawater, and on the other side you could see the town of Bowmore. Beyond that, on the horizon, you could see some of Islay's biggest mountaintops. Big is relative, as they rise only about fifteen hundred feet. But the view was no less impressive. This island would be the last stop on my journey, and it had enchanted me from the very start.

* * *

The next morning we woke to the sounds and smells of someone working in the kitchen, and we walked downstairs to find another of Bruichladdich's hospitality employees (who, coincidentally, was also named Sharon) prepping breakfast. As if she knew the exact time we would descend from our room, after sitting at the dining table (which also overlooked Loch Indaal) for no more than a minute, she emerged from the kitchen with a full Scottish breakfast. A Scottish breakfast is like an Irish breakfast, but it includes the addition of a potato scone, to which I immediately became addicted. After breakfast, we headed to the distillery to meet Christy McFarlane.

Christy is the communications manager for Bruichladdich's malt whiskies. She grew up on Islay. Her first job in whisky was with Ardbeg as a seasonal tour guide while still in high school. After school she went to the University of Edinburgh and earned a degree in business management. There she was chairman of its Water of Life Whisky Club, and from the get-go she knew she wanted to return to the whisky industry after graduation. She intended to work for a distillery on the Scottish mainland, but she said something "drew her back to Islay." She joined Bruichladdich in 2015.

Leah and I entered the Bruichladdich visitor center, and the first thing I saw was not a whisky but a gin. In 2011, Bruichladdich released the Botanist, a gin made from twenty-two locally picked wild Islay botanicals. The gin's creation might be the pinnacle of Bruichladdich's pursuit and respect for terroir. One of their slogans is "We believe terroir matters," and they express it in multiple ways: through their bottles, their website, and their distillery walls. There was a table exhibit in the visitor center showing some of the wild Islay botanicals. As I was inspecting them, Christy emerged from an office door. After exchanging introductions, she grabbed a bottle of whisky and invited us to hop in her car and go for a drive.

As we drove, Christy gave us the background of Bruichladdich. It had been built in 1881 by the Harvey brothers, William, John, and Robert. The Harvey family had operated two distilleries in Glasgow since 1770, so the brothers brought their inheritances and talents to create what was, at its opening, one of the most advanced distilleries of the day. William ran the distillery until his death in 1936, and over the next fifty years, it would change hands many times. In 1968, Invergordon Distillers bought it and ran it until 1993, when they sold to Whyte & Mackay. That lasted two years, and in 1995 the distillery was mothballed. Production resumed for a few months in 1998 before Whyte & Mackay decided to halt production officially, leaving the future of the distillery and its stock of maturing barrels unknown.

Enter Mark Reynier. Reynier was running the scotch whisky bottler Murray McDavid with Simon Coughlin. In 2000, Murray McDavid bought Bruichladdich. As they dove more into the standard production techniques for whisky, they were surprised to find just how little importance was placed on the flavor of the barley itself. Instead of flavor, distillers were more worried about how much alcohol could be yielded from the grain. Remember, Reynier's roots were in the French wine industry, so he was familiar with the importance winemakers placed on grape flavor.

"This was against everything they believed in," Christy told me. "And it wasn't just the focus on yield and efficiency over flavor. It was how the raw ingredient—barley—was treated as a commodity. Mark and Simon came from the world of French wine. So it's really no surprise that they were shocked by how little importance the industry placed on barley varieties and origins. Can you imagine if French winemakers were all of a sudden told to use generic, commodity grapes that may have been grown hundreds of miles from the winery? That's kind of what happened to Mark and Simon when they first entered the whisky industry."

Reynier and Coughlin brought a French winemaking mindset to Bruichladdich, which Christy told me was more of a "philosophy" than a science. "In the beginning," Christy continued, "they actually pushed away the scientific approach to whisky making. They wanted to embrace the artistic approach, and they wanted to establish a *connection* to the land, to the island, and to its people."

Christy said this meant foremost a commitment to Scottish barley and to aging all of their barrels on the island. It may come as a surprise, but most Islay distilleries age the majority of their whisky barrels on the mainland. It also meant they wanted to initiate an Islay barley program. While barley had been grown on the island in the past, it was mainly for feed. Reynier and Coughlin wanted to return malting barley to Islay farms.

"The program started essentially when they bought Bruichladdich. Kentraw Farm was the first to grow Islay barley for Bruichladdich, which is quite appropriate, as it sits up the hill from the distillery. In 2004, we released the first in our line of Islay barley expressions."

Lots of distilleries say "We do things the hard way" yet embrace modern technology and automation. But Bruichladdich deserves to say it. They went against the status quo with their pledge to Scottish barley, their pursuit of Islay barley, and their steadfast commitment to Islay-based barrel maturation. And what was one of the most advanced distilleries in the nineteenth century has become one of the most manual, non-computerized, old-fashioned distilleries of the twenty-first. But to call Bruichladdich outdated misses the point.

Reynier and Coughlin enlisted Jim McEwan to be the master distiller for the reincarnated Bruichladdich. McEwan is an elder statesman of not just the scotch whisky industry but of the whiskey industry at large. You may have seen him as the lead interview on the *60 Minutes* episode on Islay whisky in 2018. He joined the whiskey industry when he was just fifteen, as an apprentice cooper at Bowmore. He was born two hundred meters from the distillery, and his grandfather was a maltster there, in an era when distilleries often had a malthouse. He rose the ranks to distillery manager at Bowmore before Reynier and Coughlin hired him away.

McEwan oversaw the dismantling and reassembling of the distillery, and in late 2011, the once defunct distillery resumed operations.

Driving west across the Rinns from Port Charlotte, we could see the Atlantic Ocean whenever we crested hills. Still not totally sure where we were going, I started to wonder if maybe Christy just wanted to take a joyride to pass the time and enjoy the scenery. But before long, Christy slowed down and pulled the car up along a field of what I assumed, at first, was barley.

"So . . ." Christy said slowly. From the windup I knew she was about to reveal something unexpected. "This here is a field of winter rye."

I had come to Islay to learn about terroir in barley, and instead I had found rye.

"This is Andrew Jones's field. He runs Coull Farm here on the Rinns. Andrew is one of the youngest farmers on Islay and definitely the most curious. He likes to push boundaries: If they say it can't be done on Islay, Andrew says, 'Watch me.'"

It was a beautiful crop. Sturdy stalks supporting fat heads, with little if any signs of lodging. Christy told me this winter variety was doing better than the spring variety they had tried last year. Winter varieties fare better in colder temperatures, and they're planted so their growth phase corresponds with the colder months.

Of all the farms and places I visited that were growing rye, this was the most astonishing. On this tiny island off the western coast of Scotland, a land known solely for malt barley whisky, a farmer-distiller collaboration had successfully planted a crop that had never been grown before. This would become a completely new type of whisky. In their pursuit of terroir, Bruichladdich was pushing boundaries and expanding the diversity of flavors for their drinkers. They didn't know exactly what those flavors would be or which varieties would express which flavors through the environment of Islay. But that didn't matter—what mattered was the chase, the pursuit.

"We don't even know what we are going to call the spirit made from rye," Christy said, as we looked out to the horizon where the field of rye met the rugged Hebridean hills. "The grain bill will contain a majority of rye, complemented by malted barley. Technically, that will make it a *scotch single grain whisky* because the Scotch Whisky Regulations won't allow us to call it a rye whisky."

"How did you all convince Andrew to grow rye?" I asked Christy. Farmers are usually skeptical about planting crops that have no track record of success.

Christy laughed. "We didn't have convince him. He wanted to do it. That's just who he is. He's constantly wanting to try new things. Pardon the pun, but he likes to go against the grain."

We hopped back in the car and continued west across the Rinns. Before long, the Atlantic Ocean came into full view. The day was cloudy, and soft rains fell intermittently from a low and misty sky. We were on a paved one-lane road, but we turned off onto an even narrower gravel road before stopping at a gate.

"Put on your boots," Christy said. "We're going to take a hike." She told us we were still on Coull Farm, crossing over the grazing lands. Andrew raises about seventy head of cattle and a thousand sheep at any given time. We followed a small walking trail that was flanked by the farm's fields. There were sheep rummaging around, nibbling away at the grass. The course wavered some, but it was obvious where we were heading: the ocean.

At the edge of the farmlands the grass met a hill covered in sand-dune vegetation. Below was a long beach of white sands. Beyond that was the Atlantic Ocean, thousands of miles of it stretching away. We climbed down the hill and stumbled onto the beach. I had seen Christy grab the bottle of whisky and a couple of crystal glasses when we exited the car. It was 10 a.m., but I did not object.

The bottle was Bruichladdich's 2010 Islay Barley expression. To create it, they'd planted two different barley varieties—Optic and Oxbridge—across eight Islay farms, including Coull. They combined the harvests, distilled in 2010, and aged the new-make in 75 percent first-fill bourbon casks and 25 percent second-fill sweet wine casks, which had once held Rivesaltes, Jurançon, and Banyuls. "First-fill" means the barrels have been used only once before for maturation. "First-fill bourbon" means the barrels first aged bourbon. "Second-fill" means the barrels have already been used twice. A second-fill delivers less barrel character than a first-fill, especially when that barrel held sweet wine. Christy said the use of 25 percent second-fill was intentional. They were ensuring that the barley flavors were sufficiently maintained.

The whisky was light and floral, buffeted underneath by a sweet maltiness. I detected a slight brininess too, but maybe that was just the sea air. It was delicious. It reminded me of sharing a glass with Seamus in his farmhouse, in that it was more than just the flavor that made it an experience. Part of it was knowing I was just down the hill from where this whisky began its life as barley.

"I know this must seem a bit contrived, bringing you to a beach to drink whisky," Christy said. "But I wanted you to see that there's something about this island in how we make whisky. The sea, the weather, the character of its people—something elemental."

Bruichladdich's whole philosophy of "not worrying" about the science has evolved, Christy told us. Now they conduct research like Waterford and TX Whiskey do: growing different varieties of barley on different Islay farms with the intent to distill every variety/farm combination individually. They'll collaborate with the James Hutton Institute in Dundee for chemical analysis and possibly publish their results in a scientific journal. But whether the science yields anything revelatory still doesn't matter much to them.

"I remember a conversation with some other employees a while back," said Christy, "and a question came up: 'What if we discover that all this work to grow Islay barley doesn't even matter for flavor?' And somebody said, 'Well, so what? We're still supporting our local agriculture. We're still deepening our connection with Islay. This does go beyond flavor.'"

I thought about this as we walked down the beach, happily sipping whisky. I tend to gravitate toward the scientific approach to distillation more than to the artistic one. I'm often frustrated by obscure explanations of the origin and creation of whiskey flavor. I largely wrote this book because I wasn't satisfied with the endless debates and ambiguous, airy explanations of terroir. But when we talk about the experience of drinking whiskey, it's not just the detection of flavor compounds in our nose and tastebuds. It's also the experience of the whiskey's identity—its origin, its heritage, its history. That identity is inextricable from the people who made it and the place that made them.

Terroir is about the genetics of a variety and how the total growing environment encourages or discourages the expression of those genes that will ultimately influence the presence and concentration of flavor compounds. Or at least that's the cold hard scientific definition. But it's more than just that. At the risk of sounding "artistic," terroir is also about the farmers who raise the crop and the distillers who transform it into whiskey. Terroir has an undeniable human element, something that lab analysis can never capture. But it exists. And when everything comes together—the genetic code for flavor, the influence of the environment, the identities of its human curators—then I truly believe whiskey has the ability to move you in its own special, unrivaled way.

The whisky I was drinking on the beach wasn't necessarily better than most of the whiskeys I try. It was good—delicious, even—but that's not why I remember it. It was because the sense of terroir was so strong, and the influence of Islay was so obvious, that the whisky built a bridge between me and its home. It could have come from nowhere else. I know if and when I drink this whisky again, I'll be transported back to that beach, to the rugged landscape of Islay, and even to the people who made it. The same can be true for all whiskeys, if we capture it. Terroir is the connection between the consumer and the place.

When we got back to Bruichladdich, we toured the distillery, learning a bit about how different barley varieties from different farms act differently during milling and mashing. One of their expressions uses an heirloom variety called Bere, which they say possesses unique and intense aromas. But Bere was also very tough on their mill because of its rock-hard seed coat, so it took longer in the processing. They've also found that barley grown on Islay seems to soak up more water during mashing than malt made from mainland barley, so they achieve lower volumes of wort after lautering. Bruichladdich isn't sure why, but they suspect it's because of the salt content of Islay barley.

We headed back to An Taigh Osda for a quick lunch, and then we were off to see more of the island. We'd seen the Rinns, so Leah and I headed northeast to Port Askaig, a few thousand feet from the Isle of Jura. Jura has one distillery (Jura) and five thousand red deer. A ferry bounces back and forth between Port Askaig and Jura all day, and another goes to Kennacraig on the mainland. As we approached the town, the morning's light misting turned into a steady rain. Deciding not to try to navigate Islay roads during a rainstorm, we parked and went into a pub attached to the town's hotel. We ordered beers and struck up a conversation with the matron behind the bar. I'm not quick to tell people I work in the whiskey industry, especially when I'm off the clock at a bar. But the owner introduced herself as Marion Spears and asked me what I did.

When I told her, she said, "Well, let me shake your hand. I'm the first cousin of Jim McEwan." Islay is small, but I didn't think it was this small. To run into the cousin of the distiller who brought Bruichladdich back from the trenches was quite the coincidence. Marion had owned the Port Bar and its hotel for many years. The bar holds the oldest continual business license in all of Islay—it was established in the sixteenth century.

Sharon had told us that Jim had a vision for what he wanted Bruichladdich to be. And it turned out that the decision to not automate was more than just stubborn Luddism or nostalgic longing or some blind pursuit for flavor. It was just as much about providing jobs to Islay. Less automation meant more hands were needed to do the job. Bruichladdich, it turns out, is the largest private employer on the island.

Sometime during our conversation, I had excused myself to use the restroom. As I got up, Sharon had said, "Take a look at the picture on the wall by the bathroom door. It's the Bowmore employees decades ago. Jim's and my grandfather is in the back row." In the picture were two rows of rough-looking Scottish men. Standing in the back row was the McEwan grandfather, holding his malt rake. The Bruichladdich I experienced today owed its existence to that stern-faced man. There it was again, I thought, terroir's human element.

CONCLUSION

THE NEXT DAY, we met with Robert McEachern, who would take us back to Coull Farm to meet Andrew Jones. Like Christy, Robert grew up on Islay, left for the mainland as soon as he could, and then returned to Islay when he realized it was where he wanted to be. Robert had gone to school with Andrew, and he had to check up on him and the barley crop, so he offered to take us along.

We drove northwest again up the Rinns. Coull Farm is unique even in Islay, as it sits on a peninsula within the Rinns peninsula, bordered to the south by Machir Bay, to the north by Saligo Bay, and to the west by the Atlantic Ocean. Like most of the farms I had visited, there was no large sign to alert us we had arrived. Instead, there was a house, a barn, gravel roads, and farm animals. We got out of the car, donned our boots and jackets (the forecast called for rain again), and headed to the barn.

Andrew was inside, repairing one of his tractors. He was tall and fit, and his face was splotched with motor grease. In his jumpsuit, he looked more like an oil-field worker in Texas than a barley farmer in Scotland. He climbed down to greet us, and in what was perhaps the thickest Scottish accent I have ever encountered, he welcomed us to his farm. I told him I didn't really have an agenda and that I'd already seen his impressive rye crop but was interested in visiting some of his barley fields and hearing about his approach to farming.

Andrew grew up on Islay, the son of a farmer. His dad had grown crops and raised livestock for over half a century. For forty years, Andrew's family had run Coull Farm. Before Bruichladdich, his dad had grown barley for feed. In 2009, at Bruichladdich's request, Andrew started growing malting barley.

We returned to the car and drove with Andrew a short distance down a road flanked by two fields of barley. In the distance to the west, the Atlantic Ocean opened up to dominate the horizon. The fields were obviously in different phases of growth; one was decidedly more brown and closer to harvest than the other. I asked Andrew what was responsible for the differences.

"Well, this one on the right is a spring malting barley that has worked well in the past," Andrew said. "It's the Sassy variety, and farmers on the mainland recommended it to me." Sassy was bred by KWS, the same plant breeder responsible for the Brasetto rye variety that Sawyer Farms grows for TX Whiskey and that Woodford Reserve is experimenting with in Kentucky. Sassy is known to resist lodging, maintain low protein levels, and mature relatively quickly.

"The field on the left is something brand new," Andrew continued. "It's a field of winter barley. As far as we know, this is the first field of winter barley that has ever been grown on Islay. I decided to try ten acres of it." The winter barley heads were lined with beautiful kernels of a slightly purple hue. It was SY Venture, bred by Syngenta. I knelt down to get a closer look. The purple came from lines that ran up and down the kernel, while the base color was the more typical brownish-green. The purple lines probably come from anthocyanins synthesized in the plant's bran, specifically in the pericarp. Anthocyanins are related to carotenoids and are similarly tasteless. But just as carotenoids are the precursors of certain terpene flavor compounds, anthocyanins are potential precursors to flavor compounds, too. Given the color, I estimated the concentration of anthocyanins was higher, which potentially meant more flavor.

"Why exactly did you choose to grow SY Venture?" I asked. "And why did you want to experiment with a winter variety at all?"

Andrew smiled a bit, shrugged his shoulders, and said, "Because they told me I couldn't."

Robert laughed. "Aye," he said. "We don't know the flavors that will come from Islay-grown winter barley. But if we never try, then we'll never know."

That's the human element of terroir.

* * *

"If you go by the textbook," says Andrew, "we shouldn't even be growing barley here."[1]

Islay's soil and weather isn't exactly ideal for growing barley, at least if you measure "ideal" by yield. The soil is either sand (which has a tendency to lock in moisture when it's wet and erode too easily when dry), peat (which is constantly waterlogged), and rock (which is hard on farm equipment). Andrew tries to "work with nature," as he told me. He grows sunflowers on the borders of his barley fields, which act as a buffer zone to protect against erosion and fertilizer runoff. But that doesn't change that the soil and weather are not exactly on the barley's side.

Maybe these conditions won't lead to spectacular planting success, harvesting success, and yield, but that doesn't mean they won't create fantastic-tasting barley. One day we might discover that when grains are exposed to stressful and competitive environments, they produce more flavor compounds and their precursors. It's true for grapes in wine and for the trees from which we cut oak barrels. Why not for whiskey grains?

If Andrew Jones wasn't brought up to push the boundaries of farming, and if Bruichladdich wasn't committed to the exploration of terroir, then we would never know the flavors of whisky made from Islay-grown barley and rye. The science will reveal itself over time. Until then, we can make our own deductions.

We can detect distinct, new, and local flavor in whiskey whether the science supports it or not. Striving for that flavor, building the necessary relationships to chase it, and realizing it in a glass of whiskey—above all else, this is the essence of terroir. Its pursuit is more than the science. A roadmap might trace the chemistry and underlying effects of the genetics and the environment, but charting terroir's entire flavor genome is still probably decades away, if it's achievable at all. But the pursuit of terroir surely pulls at threads on the flavor tapestry in ways that commodity grain does not.

To remove themselves from the commodity grain system, distilleries must establish relationships with farmers. And with every farmer-distiller collaboration I encountered or experienced, what was equally as encouraging as the whiskeys was the reinvigoration of the farmer. Instead of selling to a commodity grain system that would mix their grains into an anonymous sea of others, turning it into feed, fuel, or commodity malt, these farmers could smell and taste the fruits of their labor, many for the first time ever. And this seemed to awaken in them the same sentiment: *Now that I can taste my grain, I will do what I can to make it the best, most flavorful grain possible.* And not to be forgotten, all of them were finally getting paid a premium for their skill, their expertise, and their grain. They had gone from commodity to craft.

We thanked Andrew and headed back to Bruichladdich to say our goodbyes. On that car ride, I decided it was time to accept that I had gone as far as I could with the science of terroir. I've done my best in this book to lay out the chemistry that supports the fact that terroir affects flavor in whiskey. But what these chemical data don't do is capture the human element. A whiskey is shaped by the people who make it just as much as it is by its ingredients. Origin, heritage, history—the forces that shape a people also shape their whiskeys. There's a phrase I began using as I wrote this book, which I feel sums up what I mean by the human element: *the pursuit of terroir.* This was my eureka moment. The pursuit of terroir is ultimately more meaningful than the scientific proof of it.

For TX Whiskey, the pursuit of terroir is our decision not to use commodity grain and instead to foster a collaboration with John Sawyer so that we could distill identity-preserved grain grown only on his fields. It's our work with Dr. Seth Murray at Texas A&M to research together how terroir changes the flavor of whiskey.

For the New York Distilling Company, the pursuit is creating a New York–style rye whiskey from an heirloom that was essentially lost to agriculture and doing so by building collaborations with a local farmer and a local plant breeder.

For Kings County Distillery, it is a partnership with one of the founding pioneers of organic grain in New York and a respect for the importance that such grain can have on flavor.

For Wilderness Trail Distillery, it is a commitment to using only Kentucky-grown corn, wheat, and rye, even when the latter hasn't been a typical crop in the state for over a hundred years.

For Rabbit Hole Distilling, it is a commitment to regionality. Terroir was integrated into the land far before humans drew geographical lines. There is a case to be made that terroir is more about regionality than it is about state lines.

For Waterford Distillery, it is an unwavering commitment to the pursuit of terroir, even employing a technique never seen before—single-farm distillation—and enlisting sixty local farmers to help.

For Bruichladdich, it is a commitment to Scottish barley and Islay farms. It was hiring a master distiller whose native roots stretched back generations. And it is a collaboration with a young, determined farmer who listens to what everyone says can't be done and then rolls up his sleeves and says, "Watch me."

APPENDIX 1

WHISKEY TERROIR TASTING GUIDE

ONE OF THE most meaningful ways to experience terroir in whiskey is to taste it. While I hope this book has helped you understand and appreciate the influence that terroir has on whiskey, the experience would not be complete without a "proper tasting," as Waterford's head distiller Ned Gahan insists.

In this guide, I will first lay out the approach you can take in arranging your own whiskey terroir tasting. This approach will allow you to decide how to choose whiskeys that will highlight the effect of terroir. Not all whiskeys are created with the pursuit of terroir in mind, and even though some may appear to do so on the surface, it's important to do your own research. That said, I will also provide some specific tasting flights, ones I sampled as I wrote this book. Use them as a starting point, and then chart your own adventures from there.

APPENDIX 1

ARRANGING A WHISKEY TERROIR TASTING

STEP 1: DECIDE WHICH STYLE OF WHISKEY TO TASTE

STEP 1

CATEGORIES
Bourbon whiskey
Rye whiskey
Malt whiskey
Wheat whiskey
Corn whiskey
Barley whiskey

You have to start with a whiskey style. But I don't necessarily mean you have to choose to taste scotch whiskies, or Irish whiskeys, or American whiskeys. No, what I mean is you need to choose a style based on the *grains* that were used. So, you need to choose if you want to investigate bourbon, rye whiskey, malt whiskey, wheat whiskey, corn whiskey, or barley whiskey. The former three—bourbon, rye whiskey, and malt whiskey—are the most prevalent on liquor store shelves, will provide the most brands to choose from, and were discussed at length in this book. The latter three—wheat whiskey, corn whiskey, and barley whiskey—are less prevalent, will have a relatively limited number of brands to choose from, and were not investigated to a great extent in this book. That said, each style is interesting to explore in the context of terroir.

STEP 2: SEPARATE THE SUBCATEGORIES OF THE STYLE

Once you've chosen the style you want to investigate, you now need to identify the major subcategories of the style.

WHISKEY TERROIR TASTING GUIDE

STEP 2

CATEGORIES	SUBCATEGORIES
Bourbon whiskey	
	Ryed
	Wheated
Rye whiskey	
	High
	Low
Malt whiskey	
	Peated
	Nonpeated
Wheat whiskey	
	High
	Low
Corn whiskey	
	Oak aged
	Unaged
Barley whiskey	
	Irish pot still

Bourbon

For bourbon, this would mean you need to separate whether the *small grain complement* to corn was wheat or rye. All bourbons, by law, must contain at least 51 percent corn in the grain bill, although most will contain somewhere between 65 percent to 80 percent corn. But almost all bourbons will contain a small grain complement—either rye or wheat—to the corn. Rye is the standard small grain complement, and such bourbons can be called *ryed bourbons*. (In reality, this phrase is rarely used, and bourbons with rye as the small grain complement are simply referred to as bourbon.) Wheat is less common, and such bourbons are commonly called *wheated bourbons*. The flavors delivered by rye and wheat are fairly different. Rye is much higher in the hydroxycinnamic acids that are degraded to sweet, spicy, and smoky volatile phenols during mashing, fermentation, and distillation. There are very few bourbons that are produced from 100 percent corn. If you do choose one of these 100 percent corn bourbons for your tasting, ensure that you subcategorize them with the wheated bourbons. Wheat delivers less intense flavors to a bourbon than rye, meaning that the flavors from the corn are more prominent in wheated bourbons. Bourbons produced from 100 percent corn should hypothetically be closer to wheated bourbons than traditional bourbons that incorporate rye.

Rye Whiskey

For rye whiskey, it's important to separate your choices into subcategories based on whether the grain bill was produced from an overwhelming majority of rye (85 to 100 percent) or from at or near the bare minimum (51 percent). The former can be considered the *high* subcategory, and the latter can be considered the *low* subcategory. Rye whiskeys are produced in the United States and Canada (with some rare exceptions, such as Bruichladdich's new endeavors with rye). Canadian rye whisky will always be produced from 100 percent rye, so they will always be categorized as high-rye whiskeys. With U.S. rye whiskeys, however, you will encounter both high and low rye. Many of the non-Kentucky distilleries are producing rye whiskey from a grain bill of 85 to 100 percent rye, the most popular of which is the 95% rye/5% malted barley recipe produced by MGP Ingredients in Lawrenceburg, Indiana. At TX Whiskey, we produce a 100 percent rye whiskey, made from 85% rye/15% malted rye. Most Kentucky distilleries produce rye from at or near the 51 percent rye minimum required by law. While there are insights to be gained by comparing the *high* and *low* subcategories of rye, if you are hoping to compare the influence of terroir, it is best to separate them.

It is completely fine to arrange a terroir tasting flight that involves both Canadian and U.S. rye whiskeys. Just ensure the U.S. ones are high-rye whiskeys, and also consider if the Canadian rye whisky was aged in new, charred oak casks or used casks. If you are assembling a flight of both Canadian and U.S. rye whiskeys, I would recommend that the Canadian rye whiskies you choose are matured in new (also called virgin), charred oak casks, so that their maturation will be in line with their U.S. counterparts, which must by law be matured in new, charred oak casks.

Malt Whiskey

First, I want to clarify that when I say "malt whiskey," I am referring specifically to whiskey made from 100 percent malted barley. I am aware that there are a few whiskeys made from 100 percent malted rye, malted wheat, and even malted corn, but those are standalone products that I would not recommend—interesting as they may be—for investigating the effects of terroir. Further, in the United States, there are a few examples of malt whiskeys produced from the bare minimum (51 percent) of malted barley required by law. If you wish to include U.S. malt whiskeys in your tasting flight, ensure they are indeed produced from 100 percent malted barley. Most craft distillers are producing malt whiskey from this grain bill, and in accordance with all other countries, they use the moniker "single malt" on the label. If you see the word "single malt" on a U.S. malt whiskey, it is almost surely made from 100 percent malted barley.

For malt whiskey, it is critical that you separate *peated* from *nonpeated*. Peat is sometimes used as the fuel for the kilning phase of the malting process. When it is, the smoke—full of volatile phenols that impart smoky, rubbery, meaty, and medicinal aromas—from the burning peat infuses into the malt, and its flavors are eventually translated into the whiskey. The majority of malt whiskeys are actually nonpeated, and I would recommend you start with this category for the malt whiskey category.

That said, it is possible to do a whiskey terroir tasting flight of peated whiskeys. You will most likely be choosing from a selection of scotch single-malt whiskies from the island regions of Scotland, although there are some peated single-malt whiskeys made in Japan, Ireland, India, and the United States. But be aware that there can be a wide range of phenol concentrations present in the whiskeys, because during the malting process, the quantity of peat used and the length of time spent heating the barley can vary. Given how influential phenols from peat can be for flavor, try to ensure the whiskeys you choose are somewhat conservative in their phenol concentration. Many brands of peated malt whiskey now relay the concentration (in parts per million, which is milligrams per liter) of phenols in their whiskeys. I would suggest keeping to whiskeys that are at least within 20 parts per million (often reported as "ppm") of each other.

Wheat Whiskey

Wheat whiskey should be subcategorized in the same way that rye whiskey is. There is the *high* category, which would be whiskeys made from an overwhelming majority of wheat (85 to 100 percent). And then there is the *low* category, which would be whiskeys made from at or near the bare minimum of wheat (51 percent).

The only wheat whiskeys widely available will hail from Canada or the United States, and even then your options are drastically limited compared to bourbon, rye whiskey, and malt whiskey. But if you do embark on a wheat whiskey terroir tasting, there are some easy ways to ensure you separate the *high* and *low* categories, and they once again mirror that of rye whiskey.

Canadian wheat whisky will be produced from 100 percent wheat, and many craft distilleries in the United States are producing wheat whiskey from grain bills at or near this maximum concentration. Kentucky wheat whiskeys, on the other hand, are usually produced from grain bills at or near the 51 percent bare minimum of wheat. And as with rye whiskey, if you do compare Canadian and U.S. wheat whiskeys, ensure that the Canadian wheat whiskies have been aged in new, charred oak casks, which is the cask type required by law for U.S. wheat whiskeys.

Corn Whiskey

Both the United States and Canada produce corn whiskey. In the United States, corn whiskey must be made from a minimum of 80 percent corn, and—to differentiate it from bourbon—either must forgo maturation or be matured in used casks or matured new, uncharred (that is, toasted) oak casks. In Canada, as with their rye whiskeys and wheat whiskeys, their corn whiskeys are produced from 100 percent corn, and they are typically matured in used casks, just as most of the aged U.S. corn whiskeys. So if you are comparing *oak-aged* corn whiskey, there are no concerns about mingling U.S. and Canadian whiskeys.

What must be separated, though, are *oak-aged* corn whiskeys from *unaged corn* whiskeys. The United States is the only country that allows a whiskey style to forgo maturation, and that style is corn whiskey. While unaged corn whiskeys can be tough to find, they are perhaps one of the best options for investigating the effects of terroir, as the contribution of oak cask maturation is negated.

Barley Whiskey

Barley whiskey is different from malt (barley) whiskey in that the grain bill will contain raw, unmalted barley, usually along with some malted barley. Ireland is most famous for producing this type of whiskey, and the style is actually regulated by their government. Typical grain bills for Irish pot still whiskey will use around 60 percent barley, although the regulations state only that the grain be must be made from a minimum of 30 percent malted barley and 30 percent barley. While Irish pot still whiskeys are delicious, most on the market were produced by Irish Distillers at the Midleton Distillery. So for now, this will be a difficult style to investigate in the context of terroir. That said, as more distilleries open in Ireland and produce pot still whiskey, and as other countries such as India and the United States pursue barley whiskey, it's likely there will eventually be more opportunities to investigate the effects of terroir on this whiskey style. Nevertheless, do yourself a favor and dive into some of the delicious pot still whiskeys from the Midleton Distillery, such as Redbreast, Green Spot, Yellow Spot, Red Spot, and Midleton Very Rare. Not every whiskey tasting has to compare the effects of terroir, after all!

* * *

Steps 1 and 2 are meant to establish as much commonality among your samples as possible. Essentially, you are controlling for as many variables as you can. You are seeking flavor differences from grain and, ultimately, from terroir. Elements such as diverging grain bills and cask types will skew your tasting experience away from terroir. But realize there are many other variables that will be difficult or nearly impossible to control, for example, mashing conditions, yeast strain, fermentation temperature and length, distillation technique, and maturation site. It is still best to do as much research as possible and use your best judgment to determine which brands will provide as many controls as possible. Don't see this as a deterrent! See the need to do research as an opportunity to email, call, or visit distilleries and ask as many questions as possible. It will be a practical and fun way to continue your whiskey exploration and education. And I promise you, distillers enjoy talking about the details surrounding their ingredients and processes. Nearly every day we are asked to explain what type of barrels we use, or what makes bourbon whiskey different from scotch whisky, or what gives whiskey its color. It's not every day we get to talk to a whiskey enthusiast about yeast-strain selection, the effect of temperature during fermentation, microclimates in our barrel warehouses, or reflux ratios during distillation. So when those days come along, we're thrilled. And ultimately, the more information you can gather, the better you can plan your tasting.

STEP 3: CHOOSE YOUR GLASSWARE

Ordering over the internet makes it easy to select and buy a wide variety of glassware suitable for a whiskey tasting. The most famous for whiskey is the Glencairn whisky glass. Another great option is the Copita glass. Our sensory panel at TX Whiskey uses both of these glass types. My personal favorite is the Copita, as it is stemmed; the Glencairn is not. If you want to remove the bias of color (a very good idea as you get more serious in your exploration of whiskey), you can buy both of these glass types in blue.

What you want to avoid are tumblers, highballs, and any other glass type that is wide, tall, and possesses a large headspace. These glass types are great for enjoying a casual drink, but they do not work well in tasting flights. One reason is simply space, as tumblers and highballs take up much more space than Glencairn and Copita glasses. But when it comes to analysis, the large amount of headspace in a tumbler or highball does not allow for effective concentration of the aromas.

STEP 4: NORMALIZE ALL SAMPLES TO THE SAME PERCENT ALCOHOL BY VOLUME

Not normalizing the strength of the whiskey is one of the biggest flaws in the design of tasting flights. The strength of the whiskey in the glass will play an important role in the flavors that are recognized and perceived. This is especially true with whiskeys of high %ABV (such as barrel-proof/cask-strength expressions), where the intensity of the ethanol can make it difficult to recognize other flavors.

The equation used to ensure all samples are at the identical proof is very easy:

$$\begin{gathered}(\%\text{ABV of whiskey}) \times (\text{volume of whiskey needed}) \\ = (\text{desired } \%\text{ABV}) \times (\text{desired volume in glass})\end{gathered}$$

Let's run through a few examples.

Your measuring device will dictate part of this process. My recommendation is you buy a 50 ml graduated cylinder (they range from $6 to $10). In a standard Glencairn or Copita glass, 50 ml (1.5 ounces) is a good fill volume. If you have a graduated cylinder, here is an example of how to achieve a desired %ABV and fill volume. We're going to assume that the whiskey is 60% ABV, that our desired strength is 40% ABV, and that our desired fill volume is 50 ml.

$$\begin{aligned}&(60\%\text{ ABV}) \times (\text{volume of whiskey needed}) = (40\%\text{ ABV}) \times (50\text{ ml}) \\ &\quad = (60) \times (\text{volume of whiskey needed}) = (40) \times (50) \\ &\quad = (60) \times (\text{volume of whiskey needed}) = (2000) \\ &\quad = \text{Volume of whiskey needed} = (2000/60) \\ &\quad = \text{Volume of whiskey needed} = 33.3\text{ ml}\end{aligned}$$

Now that you have this value, you can fill your 60% ABV whiskey up to the 33 ml mark on the graduated cylinder and then top off with water to 50 ml. This will give you a total volume of 50 ml at 40% ABV. Make sure you give the glass a good swirl once the diluted whiskey has been added, which will effectively mix the whiskey and water together.

Other measuring devices can be used, and the math will not change. You'll simply exchange the ml units for whatever units you are working in.

The best %ABV for nosing and tasting is up to the preference of the drinker. There is no best answer as to what strength to conduct your tasting. The best piece of guidance would be that if the whiskeys are too high in %ABV (say, above 55%), that can quickly fatigue your palate. Assessment will most likely be best in the 40% to 50% ABV range.

It's also worth noting that while it's best to normalize samples to the same strength if possible, if they vary only by a few percentage points, then it's fine to compare without bringing them to the exact same strength.

STEP 5: NOSE AND TASTE THE WHISKEYS

(This section is adapted from a chapter in my previous book, *Shots of Knowledge: The Science of Whiskey*.)

To paraphrase the Yale neuroscientist Gordon M. Shepherd, "Flavor is not in the [whiskey]; it is created from the [whiskey] by the brain." This may seem counterintuitive, but there are other examples in the sensory world where this concept would be obvious. For example, it is well known that animals do not perceive color identically. While humans perceive a red apple to be red and a green apple to be green, a dog—because of its different assortment of color-detecting cone cells in the retina—would simply see two different shades of gray. The color of an apple is not innately fixed; it is determined by the way an organism's brain deciphers the color. The same holds true for whiskey (and all drink and food, for that matter): the flavor comes from how our brain processes the chemical signals that the whiskey creates when we smell and drink it. This means that to experience the flavor of a whiskey fully, it's important that we effectively utilize our sensory capabilities.

The flavor of whiskey starts in the world of imagination. Any previous experiences you've had drinking whiskey will create a realm of flavor expectations. For example, you might be tasting a new whiskey for the first time, but if it comes from a distillery whose other expressions you are familiar with, there is a good chance you will bias your expectations based on those other expressions. And what about color? If the color of the whiskey is a dark amber, you might assume the whiskey will boast bold flavors from charred oak. But if it's a light straw color, you might expect softer notes of fruits and flowers. These biases are unavoidable unless you take steps to blind yourself to what you are tasting. That said, I wouldn't worry too much about these biases to start. If you get more serious about whiskey tasting, then you can ask friends and family to pour samples for you, as well as use blue glasses, so that you know neither the brand of whiskey being tasted nor its color.

Whether or not you are blinded to the style, brand, or color, once the whiskey is in the glass, it's time to assess its characteristics. Initially, you can give the whiskey a light swirl and assess how long it takes the "legs" that form to drop back into the glass. The longer it takes, the more likely the whiskey is either high in alcohol, has high viscosity (which can translate to a smooth mouthfeel), or both.

Flavor in general is largely a product of smell, and this is even more pronounced when drinking whiskey, as the high ethanol content can have a numbing effect on the tongue's taste receptors. Most everyone is familiar with orthonasal smelling, which is the smell that occurs when sniffing in. More simply, orthonasal smelling is the same as nosing. When you pour a glass of whiskey and move the glass toward you, the flavor compounds will escape to the mucous lining of your nose. Olfactory receptor proteins then detect these molecules, beginning a cascade of events that ultimately lead to odor perception by the brain. I like to swirl the whiskey gently in the glass before nosing. I also like to place the bottom part of the glass against the cleft above my lip and put my nose into the glass. Some people will say that both of these methods (swirling and placing your nose into the glass) are the wrong way to nose whiskey. I disagree. But the bottom line is you should nose by whatever method allows you to best experience the aromas.

After nosing, it's time to take a sip. Take a generous amount so that it coats your tongue. Move the whiskey around your mouth slowly (don't swish), breathing in through your nose and mouth for aeration, and then breathing out through your nose to activate the second type of smelling—retronasal smelling. As you move the whiskey around and breathe, the flavor compounds are volatilized and enter your nose through the back of the throat, where the nasal and oral passages meet. The whiskey will also coat the taste buds of the tongue and pharynx. Again, taste is relatively basic compared to smell. Still, as with receptors for smell, protein receptors for taste will initiate reactions that will eventually tell the brain what sensation is being experienced. Of all three modes of experiencing flavor—orthonasal smelling (nosing), retronasal smelling, and taste—retronasal smelling is the most influential in creating flavor. So that's why it is very important to keep the whiskey in your mouth for a bit as you softly move it around and breathe.

After swallowing, the sensory stimulation will continue. The whiskey coating the pharynx will continue to be carried to odor receptors through retronasal smelling. This phase is well known to whiskey drinkers. Scientists call it the *postswallowing* phase. Whiskey (and wine, beer, and other spirit) drinkers simply call it the "finish."

As the brain processes preliminary expectations, smell, and taste, central behavior systems are activated. Memories and emotions are elicited. Motivation and reward are calculated. Language allows us to describe these experiences. This last point will likely be the most difficult for you when it comes to describing a whiskey's flavors. But you will not be alone. It's tough to find the correct words to describe flavor. Indeed, from an evolutionary standpoint, the parts of the brain for flavor and language are separate. After all, flavor is really about your body deciding "is this food or drink safe?" Being able to describe the nuances of cinnamon and lavender in a grape berry doesn't really aid us from a survival-of-the-fittest standpoint. We just need to know it's not a poison berry. That said, with practice, you will become better at linking language to flavor. When possible, use your local supermarket to learn what different spices, herbs, fruits, and flowers smell like.

STEP 6: ENJOY YOUR TASTING AND HAVE FUN!

The tables in this section show some examples of tasting flights that I think do a good job highlighting the effects of terroir. I won't provide specific tasting notes—that's a journey and experience you should take without any bias or so-called expert opinion. That said, I do provide details about grain sources, so you can easily see the close connection each of these whiskeys has to the farmers (and sometimes only one farmer) who grow their ingredients.

If you are interested in tasting notes, a simple web search of the whiskeys you are tasting will provide you with numerous "expert" reviews (often with very different notes, highlighting one reason why I personally don't like tasting notes). You can use these as references, but again, I would recommend you use your own memories and vocabulary to describe what you smell and taste.

Also, I'm going to list more whiskeys per flight than you need. While three to four whiskeys are plenty for a successful flight, I didn't want to list such a small number, as I assume some of the whiskeys I'm recommending will be difficult (or near impossible) to obtain, depending on where you live and what your local liquor stores stock. So I'll list up to ten whiskeys per flight. And if you do obtain them all—or add to the lists, as I hope you do—then you can break them up into multiple flights and mix and match them in different ways.

FLIGHT 1 RYED BOURBON

WHISKEY	STATE OF DISTILLATION	CORN SOURCE	RYE SOURCE	% ABV
Wilderness Trail Small Batch	Kentucky	Caverndale Farms, Danville, Kentucky	Walnut Grove Farms, Adairville, Kentucky	50
Rabbit Hole Heigold	Kentucky	Langley Farms, Shelbyville, Kentucky	Weyerman Malt House, Germany	47.5
Woodford Reserve	Kentucky	Various Kentucky farms	Various Kentucky farms	45.2
Ironroot Harbinger	Texas	Various Texoma farms	Various Texoma farms	57.5
Still Austin	Texas	Various Texas Hill Country farms	Various Texas Hill Country farms	50.2
Coppersea Excelsior	New York	Various New York Hudson Valley farms	Various New York Hudson Valley farms	48
Woodinville	Washington	Omlin Family Farm, Quincy, Washington	Omlin Family Farm, Quincy, Washington	45
FEW	Illinois	Regional Illinois farms	Regional Illinois farms	46
Tom's Foolery Bourbon	Ohio	Tom's Foolery Estate Farm	Tom's Foolery Estate Farm	45
Far North Spirits Bodalen	Minnesota	Far North Estate Farm	Far North Estate Farm	45

FLIGHT 2 WHEATED BOURBON

WHISKEY	STATE OF DISTILLATION	CORN SOURCE	WHEAT SOURCE	% ABV
TX Bourbon	Texas	Sawyer Farms, Hillsboro, Texas	Sawyer Farms, Hillsboro, Texas	45
Garrison Brothers Small Batch	Texas	Texas Panhandle farms	Various Texas farms	47
Wilderness Trail Single Barrel	Kentucky	Caverndale Farms, Danville, Kentucky	Caverndale Farms, Danville, Kentucky	50
Maker's Mark	Kentucky	Various Kentucky farms	Peterson Farms, Loretto, Kentucky	45
Wyoming Whiskey Small Batch	Wyoming	Rageth Farms, Byron, Wyoming	Rageth Farms, Byron, Wyoming	44
Wigle Wapsie Valley	Pennsylvania	Weatherbury Farm, Avella, Pennsylvania	Various Pennsylvania farms	46
Rock Town	Arkansas	Various farms near Little Rock, Arkansas	Various Arkansas farms	46
Finger Lakes Distilling McKenzie	New York	Various Finger Lakes Region, New York, farms	Various Finger Lakes, New York, farms	45.5
Smooth Ambler Big Level	West Virginia	Turkey Creek Farms, Union, West Virginia	Turkey Creek Farms, Union, West Virginia	50

FLIGHT 3 RYE WHISKEY—HIGH

WHISKEY	REGION / COUNTRY OF DISTILLATION	RYE SOURCE	% ABV
New York Distilling Company Ragtime Rye	New York / United States	Pederson Farms, Seneca Castle, New York	45.2
Kings County Distillery Empire Rye	New York / United States	Hudson Valley Hops & Grains, Ancramdale, New York	51
Lot 40	Ontario / Canada	Various Canada farms	43
New Riff Distilling	Kentucky / United States	Northern Europe (mainly Sweden)	50
Balcones Texas Rye	Texas / United States	Various northern and western Texas farms	50
Woodinville	Washington / United States	Omlin Family Farm, Quincy, Washington	45
Dad's Hat	Pennsylvania / United States	Various Pennsylvania farms	47.5
TX Rye*	Texas / United States	Sawyer Farms, Hillsboro, Texas	TBD
Bruichladdich Rye**	Islay / Scotland	Coull Farm, Islay, Scotland	TBD

*TX Rye will be available in 2022.

**Bruichladdich Rye release date is TBD. The grain bill is only 55 percent rye, so it technically should be placed in a rye whiskey–low tasting flight. But it's such a unique rye whiskey that I wanted to include it in this list, just to make sure it wasn't overlooked.

FLIGHT 4 MALT WHISKEY—PEATED

WHISKEY	COUNTRY / REGION OF DISTILLATION	BARLEY SOURCE	PHENOLIC CONTENT (PPM)	% ABV
Bruichladdich Octomore*	Islay / Scotland	Various farms in Islay	Varies based on expression	Varies based on expression
Kilchoman 100% Islay	Islay / Scotland	Kilchoman's Estate Farm	20	50
Amrut Peated Indian Single Malt	Karnataka / India	Various farms in northwestern India	23	46
Westland Washington Peat**	Northwest Washington / United States	Various farms in northwestern Washington	TBD	TBD
Chichibu***	Saitama / Japan	Local farms near distillery	TBD	TBD

*Bruichladdich releases several new expressions of Octomore each year. No two expressions are identical, and their barley source, phenolic content, and % ABV can change. But Bruichladdich's website does a great job revealing all these details for each expression, so just do your research before choosing one.

**While Westland does have a peated malt whiskey available, at the time of writing this book, the portion of peated malted barley comes from Scotland. However, Westland has already distilled a peated malt whiskey where the barley was grown and malted in Washington, and they even went as far as to use peat from bogs in Washington. This whiskey could hit the market as early as 2021.

***To my knowledge, Chichibu is one of the only Japanese distilleries focused on growing, malting, and using Japanese barley. Most Japanese distilleries import malted barley from Europe. Chichibu is still relatively young, and they have yet to release a whisky made from 100 percent Japanese barley. But keep them on your radar for future whiskey terroir tastings.

FLIGHT 5 MALT WHISKEY—NONPEATED

WHISKEY	COUNTRY / REGION OF DISTILLATION	BARLEY SOURCE	% ABV
Waterford*	Waterford / Ireland	Various Ireland farms	Varies based on expression
Bruichladdich Islay Barley**	Islay / Scotland	Various Islay farms	50
Aberlour 12 Year Old	Speyside / Scotland	Various Speyside farms	40
Springbank Local Barley***	Campbeltown / Scotland	Various Scottish farms	Varies based on expression
Rogue Oregon Single Malt	Oregon / United States	Rogue's Barley Farm, Tygh Valley, Oregon	40
Coppersea Big Angus Green	New York / United States	Various New York Hudson Valley farms	48
Balcones High Plains	Texas / United States	Various High Plains Texas farms	52.7
Amrut Indian Single Malt	Karnataka / India	Various northwestern India farms	46
TX Single Malt****	Texas / United States	Sawyer Farms, Hillsboro, Texas	TBD
Chichibu*****	Saitama / Japan	Various Japan farms	TBD

*Any whiskey from the Waterford Distillery would be an ideal addition to a nonpeated malt whiskey terroir tasting. In fact, you could arrange a fantastic tasting just with their various expressions alone. Their Cuvée series is a mingling of single-farm-distillation whiskeys. But Waterford also has a Single Farm Origin series, which are whiskeys made solely from a single-farm distillation. So whether it's the Cuvée or Single Farm Origin series (or both), they each will highlight aspects of terroir.

**Bruichladdich releases a new expression of their Islay Barley series each year. The barley sources change with each expression, but they are always Islay farms.

***Springbank releases a new expression of their Local Barley series each year. The barley source and % ABV will change with each expression.

****TX Single Malt will be released in 2021 or 2022.

*****To my knowledge, Chichibu is one of the only Japanese distilleries focused on growing, malting, and utilizing Japanese barley. Most Japanese distilleries import malted barley from Europe. Chichibu is still relatively young, and they have yet to release a whisky made from 100 percent Japanese barley. But keep them on your radar for future whiskey terroir tastings.

APPENDIX 2

KEY TO THE ROADMAP: SOURCES FOR CHAPTER 10

1. L. Poisson and P. Schieberle, "Characterization of the Most Odor-Active Compounds in an American Bourbon Whisky by Application of the Aroma Extract Dilution Analysis," *Journal of Agricultural and Food Chemistry* 56, no. 14 (2008): 5813–19.
2. S. Vitalini et al., "The Application of Chitosan and Benzothiadiazole in Vineyard (*Vitis vinifera* L. Cv. Groppello Gentile) Changes the Aromatic Profile and Sensory Attributes of Wine," *Food Chemistry* 162 (2014): 192–205; R. Baumes et al., "Identification and Determination of Volatile Constituents in Wines from Different Vine Cultivars," *Journal of the Science of Food and Agriculture* 37, no. 9 (1986): 927–43.
3. W. Fan et al., "Identification and Quantification of Impact Aroma Compounds in 4 Nonfloral *Vitis vinifera* Varieties Grapes," *Journal of Food Science* 75, no. 1 (2010): S81–S88.
4. H. M. Bettenhausen et al., "Influence of Malt Source on Beer Chemistry, Flavor, and Flavor Stability," *Food Research International* 113 (2018): 487–504.
5. R. López et al., "Analysis of the Aroma Intensities of Volatile Compounds Released from Mild Acid Hydrolysates of Odourless Precursors Extracted from Tempranillo and Grenache Grapes Using Gas Chromatography-Olfactometry," *Food Chemistry* 88, no. 1 (2004): 95–103; D. J. Caven-Quantrill and A. J. Buglass, "Seasonal Variation of Flavour Content of English Vineyard Grapes, Determined by Stir-Bar Sorptive Extraction–Gas Chromatography–Mass Spectrometry," *Flavour and Fragrance Journal* 23, no. 4 (2008): 239–48.
6. L. Dong et al., "Characterization of Volatile Aroma Compounds in Different Brewing Barley Cultivars," *Journal of the Science of Food and Agriculture* 95, no. 5 (2015): 915–21.
7. R. Flamini, G. De Luca, and R. Di Stefano, "Changes in Carbonyl Compounds in Chardonnay and Cabernet Sauvignon Wines as a Consequence of Malolactic Fermentation," *Vitis-Geilweilerhof* 41, no. 2 (2002): 107–12.
8. Dong et al., "Characterization of Volatile Aroma Compounds in Different Brewing Barley Cultivars."
9. Vitalini et al., "The Application of Chitosan and Benzothiadiazole"; Flamini, De Luca, and Di Stefano, "Changes in Carbonyl Compounds."
10. Dong et al., "Characterization of Volatile Aroma Compounds in Different Brewing Barley Cultivars"; A.-C. J. Cramer et al., "Analysis of Volatile Compounds from Various Types of Barley Cultivars," *Journal of Agricultural and Food Chemistry* 53, no. 19 (2005): 7526–31.
11. López et al., "Analysis of the Aroma Intensities of Volatile Compounds."

12. L. Vişan, R. Dobrinoiu, and S. Dănăilă-Guidea, "The Agrobiological Study: Technological and Olfactometry of Some Vine Varieties with Biological Resistance in Southern Romania," *Agriculture and Agricultural Science Procedia* 6 (2015): 623–30.
13. Dong et al., "Characterization of Volatile Aroma Compounds in Different Brewing Barley Cultivars."
14. Vitalini et al., "The Application of Chitosan and Benzothiadiazole"; Flamini, De Luca, and Di Stefano, "Changes in Carbonyl Compounds."
15. Dong et al., "Characterization of Volatile Aroma Compounds in Different Brewing Barley Cultivars."
16. B. Jiang and Z. Zhang, "Volatile Compounds of Young Wines from Cabernet Sauvignon, Cabernet Gernischet, and Chardonnay Varieties Grown in the Loess Plateau Region of China," *Molecules* 15, no. 12 (2010): 9184–96; B. Jiang et al., "Comparison on Aroma Compounds in Cabernet Sauvignon and Merlot Wines from Four Wine Grape–Growing Regions in China," *Food Research International* 51, no. 2 (2013): 482–89; G. Cheng et al., "Comparison Between Aroma Compounds in Wines from Four *Vitis vinifera* Grape Varieties Grown in Different Shoot Positions," *Food Science and Technology* 35, no. 2 (2015): 237–46; M. Zhang et al., "Comparative Study of Aromatic Compounds in Young Red Wines from Cabernet Sauvignon, Cabernet Franc, and Cabernet Gernischet Varieties in China," *Journal of Food Science* 72, no. 5 (2007): C248–C252.
17. M. Pozo-Bayón et al., "Effect of Vineyard Yield on the Composition of Sparkling Wines Produced from the Grape Cultivar *Parellada*," *Food Chemistry* 86, no. 3 (2004): 413–19; J. L. Aleixandre et al., "Varietal Differentiation of Red Wines in the Valencian Region (Spain)," *Journal of Agricultural and Food Chemistry* 50, no. 4 (2002): 751–55.
18. Dong et al., "Characterization of Volatile Aroma Compounds in Different Brewing Barley Cultivars."
19. Vitalini et al., "The Application of Chitosan and Benzothiadiazole"; M. Vilanova et al., "Determination of Odorants in Varietal Wines from International Grape Cultivars (*Vitis vinifera*) Grown in NW Spain," *South African Journal of Enology and Viticulture* 34, no. 2 (2013): 212–22; R. López et al., "Identification of Impact Odorants of Young Red Wines Made with Merlot, Cabernet Sauvignon, and Grenache Grape Varieties: A Comparative Study," *Journal of the Science of Food and Agriculture* 79, no. 11 (1999): 1461–467.
20. Vilanova et al., "Determination of Odorants in Varietal Wines."
21. Vitalini et al., "The Application of Chitosan and Benzothiadiazole"; Jiang et al., "Comparison on Aroma Compounds in Cabernet Sauvignon and Merlot Wines"; Zhang et al., "Comparative Study of Aromatic Compounds in Young Red Wines"; López et al., "Identification of Impact Odorants of Young Red Wines"; M. Vilanova and C. Martínez, "First Study of Determination of Aromatic Compounds of Red Wine from *Vitis vinifera cv. Castanal* Grown in Galicia (NW Spain)," *European Food Research and Technology* 224, no. 4 (2007): 431–36; R. Bramley, J. Ouzman, and P. K. Boss, "Variation in Vine Vigour, Grape Yield, and Vineyard Soils and Topography as Indicators of Variation in the Chemical Composition of Grapes, Wine, and Wine Sensory Attributes," *Australian Journal of Grape and Wine Research* 17, no. 2 (2011): 217–29; J. Green et al., "Sensory and Chemical Characterisation of Sauvignon Blanc Wine: Influence of Source of Origin," *Food Research International* 44, no. 9 (2011): 2788–97; X.-j. Wang et al., "Aroma Compounds and Characteristics of Noble-Rot Wines of Chardonnay Grapes Artificially Botrytized in the Vineyard," *Food Chemistry* 226 (2017): 41–50.
22. L. Moio and P. Etievant, "Ethyl Anthranilate, Ethyl Cinnamate, 2, 3-Dihydrocinnamate, and Methyl Anthranilate: Four Important Odorants Identified in Pinot Noir Wines of Burgundy," *American Journal of Enology and Viticulture* 46, no. 3 (1995): 392–98; Y. Fang and M. C. Qian, "Quantification of Selected Aroma-Active Compounds in Pinot Noir Wines from Different Grape Maturities," *Journal of Agricultural and Food Chemistry* 54, no. 22 (2006): 8567–73; V. Ferreira, R. López, and J. F. Cacho, "Quantitative Determination of the Odorants of Young Red Wines from Different Grape Varieties," *Journal of the Science of Food and Agriculture* 80, no. 11 (2000): 1659–67; G. Antalick et al., "Influence of Grape Composition on Red Wine

Ester Profile: Comparison Between Cabernet Sauvignon and Shiraz Cultivars from Australian Warm Climate," *Journal of Agricultural and Food Chemistry* 63, no. 18 (2015): 4664–72.

23. Vitalini et al., "The Application of Chitosan and Benzothiadiazole"; Jiang et al., "Comparison on Aroma Compounds in Cabernet Sauvignon and Merlot Wines"; Cheng et al., "Comparison Between Aroma Compounds"; Zhang et al., "Comparative Study of Aromatic Compounds in Young Red Wines"; López et al., "Identification of Impact Odorants of Young Red Wines"; J. Green et al., "Sensory and Chemical Characterisation of Sauvignon Blanc Wine: Influence of Source of Origin," *Food Research International* 44, no. 9 (2011): 2788–97; Wang et al., "Aroma Compounds and Characteristics of Noble-Rot Wines"; J. S. Câmara, M. A. Alves, and J. C. Marques, "Multivariate Analysis for the Classification and Differentiation of Madeira Wines According to Main Grape Varieties," *Talanta* 68, no. 5 (2006): 1512–21; H. Guth, "Comparison of Different White Wine Varieties in Odor Profiles by Instrumental Analysis and Sensory Studies," in *Chemistry of Wine Flavor*, ed. A. L. Waterhouse and S. E. Ebeler (Davis, CA: ACS, 1998), 39–52; Z.-m. Xi et al., "Impact of Cover Crops in Vineyard on the Aroma Compounds of *Vitis vinifera* L. Cv. Cabernet Sauvignon Wine," *Food Chemistry* 127, no. 2 (2011): 516–22.
24. Bramley, Ouzman, Boss, "Variation in Vine Vigour, Grape Yield, and Vineyard Soils and Topography"; Y. Fang and M. C. Qian, "Quantification of Selected Aroma-Active Compounds in Pinot Noir Wines from Different Grape Maturities," *Journal of Agricultural and Food Chemistry* 54, no. 22 (2006): 8567–73; Guth, "Comparison of Different White Wine Varieties"; Y. Fang and M. C. Qian, "Aroma Compounds in Oregon Pinot Noir Wine Determined by Aroma Extract Dilution Analysis (AEDA)," *Flavour and Fragrance Journal* 20, no. 1 (2005): 22–29.
25. Vitalini et al., "The Application of Chitosan and Benzothiadiazole"; Bramley, Ouzman, and Boss, "Variation in Vine Vigour, Grape Yield, and Vineyard Soils and Topography"; Wang et al., "Aroma Compounds and Characteristics of Noble-Rot Wines"; Fang and Qian, "Aroma Compounds in Oregon Pinot Noir Wine."
26. Vitalini et al., "The Application of Chitosan and Benzothiadiazole"; Jiang and Zhang, "Volatile Compounds of Young Wines"; Zhang et al., "Comparative Study of Aromatic Compounds in Young Red Wines"; López et al., "Identification of Impact Odorants of Young Red Wines"; Wang et al., "Aroma Compounds and Characteristics of Noble-Rot Wines"; Xi et al., "Impact of Cover Crops in Vineyard."
27. López et al., "Identification of Impact Odorants of Young Red Wines"; Bramley, Ouzman, and Boss, "Variation in Vine Vigour, Grape Yield, and Vineyard Soils and Topography"; Green et al., "Sensory and Chemical Characterisation of Sauvignon Blanc Wine."
28. Vitalini et al., "The Application of Chitosan and Benzothiadiazole"; J. L. Aleixandre et al., "Varietal Differentiation of Red Wines in the Valencian Region (Spain)," *Journal of Agricultural and Food Chemistry* 50, no. 4 (2002): 751–55; I. Alvarez et al., "Geographical Differentiation of White Wines from Three Subzones of the Designation of Origin Valencia," *European Food Research and Technology* 217, no. 2 (2003): 173–79.
29. Vitalini et al., "The Application of Chitosan and Benzothiadiazole"; Zhang et al., "Comparative Study of Aromatic Compounds in Young Red Wines"; Bramley, Ouzman, and Boss, "Variation in Vine Vigour, Grape Yield, and Vineyard Soils and Topography"; Green et al., "Sensory and Chemical Characterisation of Sauvignon Blanc Wine"; Xi et al., "Impact of Cover Crops in Vineyard"; E. Falqué, E. Fernández, and D. Dubourdieu, "Differentiation of White Wines by Their Aromatic Index," *Talanta* 54, no. 2 (2001): 271–81.
30. Zhang et al., "Comparative Study of Aromatic Compounds in Young Red Wines"; Bramley, Ouzman, and Boss, "Variation in Vine Vigour, Grape Yield, and Vineyard Soils and Topography"; Wang et al., "Aroma Compounds and Characteristics of Noble-Rot Wines"; Xi et al., "Impact of Cover Crops in Vineyard."
31. H. M. Bettenhausen et al., "Variation in Sensory Attributes and Volatile Compounds in Beers Brewed from Genetically Distinct Malts: An Integrated Sensory and Non-Targeted Metabolomics Approach," *Journal of the American Society of Brewing Chemists* (2020): 1–17.
32. Vitalini et al., "The Application of Chitosan and Benzothiadiazole"; Zhang et al., "Comparative Study of Aromatic Compounds in Young Red Wines"; Bramley, Ouzman, and Boss, "Variation

in Vine Vigour, Grape Yield, and Vineyard Soils and Topography"; Green et al., "Sensory and Chemical Characterisation of Sauvignon Blanc Wine"; Wang et al., "Aroma Compounds and Characteristics of Noble-Rot Wines"; Xi et al., "Impact of Cover Crops in Vineyard."

33. Dong et al., "Characterization of Volatile Aroma Compounds in Different Brewing Barley Cultivars."
34. Vitalini et al., "The Application of Chitosan and Benzothiadiazole"; Zhang et al., "Comparative Study of Aromatic Compounds in Young Red Wines"; Green et al., "Sensory and Chemical Characterisation of Sauvignon Blanc Wine"; Xi et al., "Impact of Cover Crops in Vineyard."
35. Dong et al., "Characterization of Volatile Aroma Compounds in Different Brewing Barley Cultivars."
36. Zhang et al., "Comparative Study of Aromatic Compounds in Young Red Wines"; Wang et al., "Aroma Compounds and Characteristics of Noble-Rot Wines"; I. Arozarena et al., "Multivariate Differentiation of Spanish Red Wines According to Region and Variety," *Journal of the Science of Food and Agriculture* 80, no. 13 (2000): 1909–17; A. Bellincontro et al., "Feasibility of an Electronic Nose to Differentiate Commercial Spanish Wines Elaborated from the Same Grape Variety," *Food Research International* 51, no. 2 (2013): 790–96.
37. Dong et al., "Characterization of Volatile Aroma Compounds in Different Brewing Barley Cultivars."
38. Zhang et al., "Comparative Study of Aromatic Compounds in Young Red Wines"; Bramley, Ouzman, and Boss, "Variation in Vine Vigour, Grape Yield, and Vineyard Soils and Topography"; Green et al., "Sensory and Chemical Characterisation of Sauvignon Blanc Wine"; Wang et al., "Aroma Compounds and Characteristics of Noble-Rot Wines."
39. A. L. Robinson et al., "Influence of Geographic Origin on the Sensory Characteristics and Wine Composition of *Vitis vinifera cv.* Cabernet Sauvignon Wines from Australia," *American Journal of Enology and Viticulture* 63, no. 4 (2012): 467–76.
40. H. Guth, "Identification of Character Impact Odorants of Different White Wine Varieties," *Journal of Agricultural and Food Chemistry* 45, no. 8 (1997): 3022–26; H. Guth, "Quantitation and Sensory Studies of Character Impact Odorants of Different White Wine Varieties," *Journal of Agricultural and Food Chemistry* 45, no. 8 (1997): 3027–32.
41. Dong et al., "Characterization of Volatile Aroma Compounds in Different Brewing Barley Cultivars."
42. Guth, "Quantitation and Sensory Studies"; T. E. Siebert et al., "Analysis, Potency, and Occurrence of (Z)-6-Dodeceno-γ-Lactone in White Wine," *Food Chemistry* 256 (2018): 85–90.
43. Wang et al., "Aroma Compounds and Characteristics of Noble-Rot Wines"; R. C. Cooke et al., "Quantification of Several 4-Alkyl Substituted γ-lactones in Australian Wines," *Journal of Agricultural and Food Chemistry* 57, no. 2 (2009): 348–52; J. Langen et al., "Quantitative Analysis of γ- and δ-lactones in Wines Using Gas Chromatography with Selective Tandem Mass Spectrometric Detection," *Rapid Communications in Mass Spectrometry* 27, no. 24 (2013): 2751–59.
44. López et al., "Analysis of the Aroma Intensities of Volatile Compounds"; Langen et al., "Quantitative Analysis of γ- and δ-lactones in Wines."
45. Wang et al., "Aroma Compounds and Characteristics of Noble-Rot Wines"; Ferreira, López, and Cacho, "Quantitative Determination of the Odorants of Young Red Wines"; Langen et al., "Quantitative Analysis of γ- and δ-lactones in Wines"; S. Nakamura et al., "Quantitative Analysis of γ-nonalactone in Wines and Its Threshold Determination," *Journal of Food Science* 53, no. 4 (1988): 1243–44.
46. Langen et al., "Quantitative Analysis of γ- and δ-lactones in Wines"; O. Vyviurska and I. Špánik, "Assessment of Tokaj Varietal Wines with Comprehensive Two-Dimensional Gas Chromatography Coupled to High Resolution Mass Spectrometry," *Microchemical Journal* 152 (2020): 104385.
47. M. J. Lacey et al., "Methoxypyrazines in Sauvignon Blanc Grapes and Wines," *American Journal of Enology and Viticulture* 42, no. 2 (1991): 103–8; G. Pickering et al., "Determination of Ortho- and Retronasal Detection Thresholds for 2-Isopropyl-3-Methoxypyrazine in Wine," *Journal of Food Science* 72, no. 7 (2007): S468–S472; D. Sidhu et al., "Methoxypyrazine

Analysis and Influence of Viticultural and Enological Procedures on Their Levels in Grapes, Musts, and Wines," *Critical Reviews in Food Science and Nutrition* 55, no. 4 (2015): 485–502.

48. López et al., "Analysis of the Aroma Intensities of Volatile Compounds"; Wang et al., "Aroma Compounds and Characteristics of Noble-Rot Wines"; A. Buettner, "Investigation of Potent Odorants and Afterodor Development in Two Chardonnay Wines Using the Buccal Odor Screening System (BOSS)," *Journal of Agricultural and Food Chemistry* 52, no. 8 (2004): 2339–46.
49. Vitalini et al., "The Application of Chitosan and Benzothiadiazole"; Zhang et al., "Comparative Study of Aromatic Compounds in Young Red Wines"; López et al., "Identification of Impact Odorants of Young Red Wines"; Bramley, Ouzman, and Boss, "Variation in Vine Vigour, Grape Yield, and Vineyard Soils and Topography"; Wang et al., "Aroma Compounds and Characteristics of Noble-Rot Wines"; Xi et al., "Impact of Cover Crops in Vineyard"; C. Ou et al., "Volatile Compounds and Sensory Attributes of Wine from cv. Merlot (*Vitis vinifera* L.) Grown Under Differential Levels of Water Deficit with or Without a Kaolin-Based, Foliar Reflectant Particle Film," *Journal of Agricultural and Food Chemistry* 58, no. 24 (2010): 12890–898; S.-H. Lee et al., "Vine Microclimate and Norisoprenoid Concentration in Cabernet Sauvignon Grapes and Wines," *American Journal of Enology and Viticulture* 58, no. 3 (2007): 291–301; M. C. Qian, Y. Fang, and K. Shellie, "Volatile Composition of Merlot Wine from Different Vine Water Status," *Journal of Agricultural and Food Chemistry* 57, no. 16 (2009): 7459–63.
50. López et al., "Identification of Impact Odorants of Young Red Wines"; Wang et al., "Aroma Compounds and Characteristics of Noble-Rot Wines"; J. Song et al., "Pinot Noir Wine Composition from Different Vine Vigour Zones Classified by Remote Imaging Technology," *Food Chemistry* 153 (2014): 52–59; Y. Kotseridis et al., "Quantitative Determination of β-ionone in Red Wines and Grapes of Bordeaux Using a Stable Isotope Dilution Assay," *Journal of Chromatography A* 848, no. 1–2 (1999): 317–25; J. d. S. Câmara et al., "Varietal Flavour Compounds of Four Grape Varieties Producing Madeira Wines," *Analytica Chimica Acta* 513, no. 1 (2004): 203–7.
51. M. A. Segurel et al., "Contribution of Dimethyl Sulfide to the Aroma of Syrah and Grenache Noir Wines and Estimation of Its Potential in Grapes of These Varieties," *Journal of Agricultural and Food Chemistry* 52, no. 23 (2004): 7084–93; B. Fedrizzi et al., "Aging Effects and Grape Variety Dependence on the Content of Sulfur Volatiles in Wine," *Journal of Agricultural and Food Chemistry* 55, no. 26 (2007): 10880–87; S. K. Park et al., "Incidence of Volatile Sulfur Compounds in California Wines: A Preliminary Survey," *American Journal of Enology and Viticulture* 45, no. 3 (1994): 341–44.
52. B. Yang, P. Schwarz, and R. Horsley, "Factors Involved in the Formation of Two Precursors of Dimethylsulfide During Malting," *Journal of the American Society of Brewing Chemists* 56, no. 3 (1998): 85–92; C. W. Bamforth, "Dimethyl Sulfide—Significance, Origins, and Control," *Journal of the American Society of Brewing Chemists* 72, no. 3 (2014): 165–68.
53. López et al., "Identification of Impact Odorants of Young Red Wines"; Qian, Fang, and Shellie, "Volatile Composition of Merlot Wine"; E. Gómez-Plaza et al., "Investigation on the Aroma of Wines from Seven Clones of Monastrell Grapes," *European Food Research and Technology* 209, no. 3–4 (1999): 257–60; E. G. García-Carpintero et al., "Volatile and Sensory Characterization of Red Wines from cv. Moravia Agria Minority Grape Variety Cultivated in La Mancha Region Over Five Consecutive Vintages," *Food Research International* 44, no. 5 (2011): 1549–60.
54. López et al., "Identification of Impact Odorants of Young Red Wines"; Qian, Fang, and Shellie, "Volatile Composition of Merlot Wine"; Gómez-Plaza et al., "Investigation on the Aroma of Wines"; García-Carpintero et al., "Volatile and Sensory Characterization of Red Wines."
55. López et al., "Identification of Impact Odorants of Young Red Wines"; Wang et al., "Aroma Compounds and Characteristics of Noble-Rot Wines"; García-Carpintero et al., "Volatile and Sensory Characterization of Red Wines."
56. López et al., "Identification of Impact Odorants of Young Red Wines"; Wang et al., "Aroma Compounds and Characteristics of Noble-Rot Wines"; Qian, Fang, and Shellie, "Volatile Composition of Merlot Wine"; García-Carpintero et al., "Volatile and Sensory Characterization of Red Wines."

APPENDIX 3

KEY TO THE ROADMAP: SOURCES FOR CHAPTER 11

1. L. Poisson and P. Schieberle, "Characterization of the Most Odor-Active Compounds in an American Bourbon Whisky by Application of the Aroma Extract Dilution Analysis," *Journal of Agricultural and Food Chemistry* 56, no. 14 (2008): 5813–19.
2. J. Lahne, "Aroma Characterization of American Rye Whiskey by Chemical and Sensory Assays," master's thesis, University of Illinois at Urbana-Champaign, 2010, http://hdl.handle.net/2142/16713.
3. S. Vitalini et al., "The Application of Chitosan and Benzothiadiazole in Vineyard (*Vitis vinifera* L. Cv. Groppello Gentile) Changes the Aromatic Profile and Sensory Attributes of Wine," *Food Chemistry* 162 (2014): 192–205; R. Baumes et al., "Identification and Determination of Volatile Constituents in Wines from Different Vine Cultivars," *Journal of the Science of Food and Agriculture* 37, no. 9 (1986): 927–43.
4. W. Fan et al., "Identification and Quantification of Impact Aroma Compounds in 4 Nonfloral *Vitis vinifera* Varieties Grapes," *Journal of Food Science* 75, no. 1 (2010): S81–S88.
5. H. M. Bettenhausen et al., "Influence of Malt Source on Beer Chemistry, Flavor, and Flavor Stability," *Food Research International* 113 (2018): 487–504.
6. R. López et al., "Mild Acid Hydrolysates of Odourless Precursors Extracted from Tempranillo and Grenache Grapes Using Gas Chromatography-Olfactometry," *Food Chemistry* 88, no. 1 (2004): 95–103; D. J. Caven-Quantrill and A. J. Buglass, "Seasonal Variation of Flavour Content of English Vineyard Grapes, Determined by Stir-Bar Sorptive Extraction–Gas Chromatography–Mass Spectrometry," *Flavour and Fragrance Journal* 23, no. 4 (2008): 239–48.
7. L. Dong et al., "Characterization of Volatile Aroma Compounds in Different Brewing Barley Cultivars," *Journal of the Science of Food and Agriculture* 95, no. 5 (2015): 915–21.
8. R. Flamini, G. De Luca, and R. Di Stefano, "Changes in Carbonyl Compounds in Chardonnay and Cabernet Sauvignon Wines as a Consequence of Malolactic Fermentation," *Vitis-Geilweilerhof* 41, no. 2 (2002): 107–12.
9. Dong et al., "Characterization of Volatile Aroma Compounds."
10. Vitalini et al., "The Application of Chitosan and Benzothiadiazole"; Flamini, De Luca, and Di Stefano, "Changes in Carbonyl Compounds."
11. Dong et al., "Characterization of Volatile Aroma Compounds"; A.-C. J. Cramer et al., "Analysis of Volatile Compounds from Various Types of Barley Cultivars," *Journal of Agricultural and Food Chemistry* 53, no. 19 (2005): 7526–31.

12. R. López et al., "Analysis of the Aroma Intensities of Volatile Compounds Released from Mild Acid Hydrolysates of Odourless Precursors Extracted from Tempranillo and Grenache Grapes Using Gas Chromatography-Olfactometry," *Food Chemistry* 88, no. 1 (2004): 95–103.
13. L. Vişan, R. Dobrinoiu, and S. Dănăilă-Guidea, "The Agrobiological Study: Technological and Olfactometry of Some Vine Varieties with Biological Resistance in Southern Romania," *Agriculture and Agricultural Science Procedia* 6 (2015): 623–30.
14. Dong et al., "Characterization of Volatile Aroma Compounds."
15. Vitalini et al., "The Application of Chitosan and Benzothiadiazole"; Flamini, De Luca, and Di Stefano, "Changes in Carbonyl Compounds."
16. Dong et al., "Characterization of Volatile Aroma Compounds."
17. B. Jiang and Z. Zhang, "Volatile Compounds of Young Wines from Cabernet Sauvignon, Cabernet Gernischet, and Chardonnay Varieties Grown in the Loess Plateau Region of China," *Molecules* 15, no. 12 (2010): 9184–96; B. Jiang et al., "Comparison of Aroma Compounds in Cabernet Sauvignon and Merlot Wines from Four Wine Grape–Growing Regions in China," *Food Research International* 51, no. 2 (2013): 482–89; G. Cheng et al., "Comparison Between Aroma Compounds in Wines from Four *Vitis vinifera* Grape Varieties Grown in Different Shoot Positions," *Food Science and Technology* 35, no. 2 (2015): 237–46; M. Zhang et al., "Comparative Study of Aromatic Compounds in Young Red Wines from Cabernet Sauvignon, Cabernet Franc, and Cabernet Gernischet Varieties in China," *Journal of Food Science* 72, no. 5 (2007): C248–C252.
18. M. Pozo-Bayón et al., "Effect of Vineyard Yield on the Composition of Sparkling Wines Produced from the Grape Cultivar *Parellada*," *Food Chemistry* 86, no. 3 (2004): 413–19; J. L. Aleixandre et al., "Varietal Differentiation of Red Wines in the Valencian Region (Spain)," *Journal of Agricultural and Food Chemistry* 50, no. 4 (2002): 751–55.
19. Dong et al., "Characterization of Volatile Aroma Compounds."
20. Vitalini et al., "The Application of Chitosan and Benzothiadiazole"; M. Vilanova et al., "Determination of Odorants in Varietal Wines from International Grape Cultivars (*Vitis vinifera*) Grown in NW Spain," *South African Journal of Enology and Viticulture* 34, no. 2 (2013): 212–22; R. López et al., "Identification of Impact Odorants of Young Red Wines Made with Merlot, Cabernet Sauvignon, and Grenache Grape Varieties: A Comparative Study," *Journal of the Science of Food and Agriculture* 79, no. 11 (1999): 1461–67.
21. Zhang et al., "Comparative Study of Aromatic Compounds in Young Red Wines."
22. Vitalini et al., "The Application of Chitosan and Benzothiadiazole"; Jiang et al., "Comparison of Aroma Compounds in Cabernet Sauvignon and Merlot Wines"; Zhang et al., "Comparative Study of Aromatic Compounds in Young Red Wines"; López et al., "Identification of Impact Odorants of Young Red Wines"; M. Vilanova and C. Martínez, "First Study of Determination of Aromatic Compounds of Red Wine from *Vitis vinifera cv. Castanal* Grown in Galicia (NW Spain)," *European Food Research and Technology* 224, no. 4 (2007): 431–36; R. Bramley, J. Ouzman, and P. K. Boss, "Variation in Vine Vigour, Grape Yield, and Vineyard Soils and Topography as Indicators of Variation in the Chemical Composition of Grapes, Wine, and Wine Sensory Attributes," *Australian Journal of Grape and Wine Research* 17, no. 2 (2011): 217–29; J. Green et al., "Sensory and Chemical Characterisation of Sauvignon Blanc Wine: Influence of Source of Origin," *Food Research International* 44, no. 9 (2011): 2788–97; X.-j. Wang et al., "Aroma Compounds and Characteristics of Noble-Rot Wines of Chardonnay Grapes Artificially Botrytized in the Vineyard," *Food Chemistry* 226 (2017): 41–50.
23. L. Moio and P. Etievant, "Ethyl Anthranilate, Ethyl Cinnamate, 2, 3-Dihydrocinnamate, and Methyl Anthranilate: Four Important Odorants Identified in Pinot Noir Wines of Burgundy," *American Journal of Enology and Viticulture* 46, no. 3 (1995): 392–98; Y. Fang and M. C. Qian, "Quantification of Selected Aroma-Active Compounds in Pinot Noir Wines from Different Grape Maturities," *Journal of Agricultural and Food Chemistry* 54, no. 22 (2006): 8567–73; V. Ferreira, R. López, and J. F. Cacho, "Quantitative Determination of the Odorants of Young Red Wines from Different Grape Varieties," *Journal of the Science of Food and Agriculture* 80, no. 11 (2000): 1659–67; G. Antalick et al., "Influence of Grape Composition on Red Wine

Ester Profile: Comparison Between Cabernet Sauvignon and Shiraz Cultivars from Australian Warm Climate," *Journal of Agricultural and Food Chemistry* 63, no. 18 (2015): 4664–72.

24. Vitalini et al., "The Application of Chitosan and Benzothiadiazole"; Jiang et al., "Comparison of Aroma Compounds in Cabernet Sauvignon and Merlot Wines"; Cheng et al., "Comparison Between Aroma Compounds"; Zhang et al., "Comparative Study of Aromatic Compounds in Young Red Wines"; López et al., "Identification of Impact Odorants of Young Red Wines"; Green et al., "Sensory and Chemical Characterisation of Sauvignon Blanc Wine"; Wang et al., "Aroma Compounds and Characteristics of Noble-Rot Wines"; J. S. Câmara, M. A. Alves, and J. C. Marques, "Multivariate Analysis for the Classification and Differentiation of Madeira Wines According to Main Grape Varieties," *Talanta* 68, no. 5 (2006): 1512–21; H. Guth, "Comparison of Different White Wine Varieties in Odor Profiles by Instrumental Analysis and Sensory Studies," in *Chemistry of Wine Flavor*, ed. A. L. Waterhouse and S. E. Ebeler (Davis, CA: ACS, 1998), 39–52; Z.-m. Xi et al., "Impact of Cover Crops in Vineyard on the Aroma Compounds of *Vitis vinifera* L. Cv. Cabernet Sauvignon Wine," *Food Chemistry* 127, no. 2 (2011): 516–22.
25. Bramley, Ouzman, and Boss, "Variation in Vine Vigour, Grape Yield, and Vineyard Soils and Topography"; Fang and Qian, "Quantification of Selected Aroma-Active Compounds in Pinot Noir Wines"; Guth, "Comparison of Different White Wine Varieties in Odor Profiles"; Y. Fang and M. C. Qian, "Aroma Compounds in Oregon Pinot Noir Wine Determined by Aroma Extract Dilution Analysis (AEDA)," *Flavour and Fragrance Journal* 20, no. 1 (2005): 22–29.
26. Vitalini et al., "The Application of Chitosan and Benzothiadiazole"; Bramley, Ouzman, and Boss, "Variation in Vine Vigour, Grape Yield, and Vineyard Soils and Topography"; Wang et al., "Aroma Compounds and Characteristics of Noble-Rot Wines"; Fang and Qian, "Aroma Compounds in Oregon Pinot Noir Wine Determined by Aroma Extract Dilution Analysis (AEDA)."
27. Vitalini et al., "The Application of Chitosan and Benzothiadiazole"; Jiang and Zhang, "Volatile Compounds of Young Wines from Cabernet Sauvignon, Cabernet Gernischet, and Chardonnay Varieties Grown in the Loess Plateau Region of China"; Zhang et al., "Comparative Study of Aromatic Compounds in Young Red Wines from Cabernet Sauvignon, Cabernet Franc, and Cabernet Gernischet Varieties in China"; R. López et al., "Identification of Impact Odorants of Young Red Wines Made with Merlot, Cabernet Sauvignon, and Grenache Grape Varieties"; Xi et al., "Impact of Cover Crops."
28. R. López et al., "Identification of Impact Odorants of Young Red Wines Made with Merlot, Cabernet Sauvignon, and Grenache Grape Varieties"; Bramley, Ouzman, and Boss, "Variation in Vine Vigour, Grape Yield, and Vineyard Soils and Topography"; Green et al., "Sensory and Chemical Characterisation of Sauvignon Blanc Wine."
29. Vitalini et al., "The Application of Chitosan and Benzothiadiazole"; Aleixandre et al., "Varietal Differentiation of Red Wines in the Valencian Region (Spain)"; I. Alvarez et al., "Geographical Differentiation of White Wines from Three Subzones of the Designation of Origin Valencia," *European Food Research and Technology* 217, no. 2 (2003): 173–79.
30. Vitalini et al., "The Application of Chitosan and Benzothiadiazole"; Zhang et al., "Comparative Study of Aromatic Compounds in Young Red Wines from Cabernet Sauvignon, Cabernet Franc, and Cabernet Gernischet Varieties in China"; Bramley, Ouzman, and Boss, "Variation in Vine Vigour, Grape Yield, and Vineyard Soils and Topography"; Green et al., "Sensory and Chemical Characterisation of Sauvignon Blanc Wine"; Xi et al., "Impact of Cover Crops"; E. Falqué, E. Fernández, and D. Dubourdieu, "Differentiation of White Wines by Their Aromatic Index," *Talanta* 54, no. 2 (2001): 271–81.
31. Zhang et al., "Comparative Study of Aromatic Compounds in Young Red Wines from Cabernet Sauvignon, Cabernet Franc, and Cabernet Gernischet Varieties in China"; Bramley, Ouzman, and Boss, "Variation in Vine Vigour, Grape Yield, and Vineyard Soils and Topography"; Wang et al., "Aroma Compounds and Characteristics of Noble-Rot Wines"; Xi et al., "Impact of Cover Crops."

32. H. M. Bettenhausen et al., "Variation in Sensory Attributes and Volatile Compounds in Beers Brewed from Genetically Distinct Malts: An Integrated Sensory and Non-Targeted Metabolomics Approach," *Journal of the American Society of Brewing Chemists* (2020): 1–17.
33. Vitalini et al., "The Application of Chitosan and Benzothiadiazole"; Zhang et al., "Comparative Study of Aromatic Compounds in Young Red Wines"; Bramley, Ouzman, and Boss, "Variation in Vine Vigour, Grape Yield, and Vineyard Soils and Topography"; Green et al., "Sensory and Chemical Characterisation of Sauvignon Blanc Wine"; Wang et al., "Aroma Compounds and Characteristics of Noble-Rot Wines"; Xi et al., "Impact of Cover Crops."
34. Dong et al., "Characterization of Volatile Aroma Compounds."
35. S. Pérez-Magariño et al., "Multivariate Analysis for the Differentiation of Sparkling Wines Elaborated from Autochthonous Spanish Grape Varieties: Volatile Compounds, Amino Acids, and Biogenic Amines," *European Food Research and Technology* 236, no. 5 (2013): 827–41; O. Martínez-Pinilla et al., "Characterization of Volatile Compounds and Olfactory Profile of Red Minority Varietal Wines from La Rioja," *Journal of the Science of Food and Agriculture* 93, no. 15 (2013): 3720–29. ; M. Vilanova et al., "Volatile Composition and Sensory Properties of North West Spain White Wines," *Food Research International* 54, no. 1 (2013): 562–68; A. Slegers et al., "Volatile Compounds from Grape Skin, Juice, and Wine from Five Interspecific Hybrid Grape Cultivars Grown in Québec (Canada) for Wine Production," *Molecules* 20, no. 6 (2015): 10980–1016.
36. Zhang et al., "Comparative Study of Aromatic Compounds in Young Red Wines"; Bramley, Ouzman, and Boss, "Variation in Vine Vigour, Grape Yield, and Vineyard Soils and Topography"; Green et al., "Sensory and Chemical Characterisation of Sauvignon Blanc Wine"; Xi et al., "Impact of Cover Crops."
37. Dong et al., "Characterization of Volatile Aroma Compounds."
38. Zhang et al., "Comparative Study of Aromatic Compounds in Young Red Wines"; Wang et al., "Aroma Compounds and Characteristics of Noble-Rot Wines"; I. Arozarena et al., "Multivariate Differentiation of Spanish Red Wines According to Region and Variety," *Journal of the Science of Food and Agriculture* 80, no. 13 (2000): 1909–17; A. Bellincontro et al., "Feasibility of an Electronic Nose to Differentiate Commercial Spanish Wines Elaborated from the Same Grape Variety," *Food Research International* 51, no. 2 (2013): 790–96.
39. Dong et al., "Characterization of Volatile Aroma Compounds."
40. Zhang et al., "Comparative Study of Aromatic Compounds in Young Red Wines"; Bramley, Ouzman, and Boss, "Variation in Vine Vigour, Grape Yield, and Vineyard Soils and Topography"; Green et al., "Sensory and Chemical Characterisation of Sauvignon Blanc Wine"; Wang et al., "Aroma Compounds and Characteristics of Noble-Rot Wines."
41. A. L. Robinson et al., "Influence of Geographic Origin on the Sensory Characteristics and Wine Composition of *Vitis vinifera cv.* Cabernet Sauvignon Wines from Australia," *American Journal of Enology and Viticulture* 63, no. 4 (2012): 467–76.
42. H. Guth, "Identification of Character Impact Odorants of Different White Wine Varieties," *Journal of Agricultural and Food Chemistry* 45, no. 8 (1997): 3022–26; H. Guth, "Quantitation and Sensory Studies of Character Impact Odorants of Different White Wine Varieties," *Journal of Agricultural and Food Chemistry* 45, no. 8 (1997): 3027–32.
43. Dong et al., "Characterization of Volatile Aroma Compounds."
44. Guth, "Quantitation and Sensory Studies of Character Impact Odorants of Different White Wine Varieties"; T. E. Siebert et al., "Analysis, Potency, and Occurrence of (Z)-6-Dodeceno-γ-Lactone in White Wine," *Food Chemistry* 256 (2018): 85–90.
45. Wang et al., "Aroma Compounds and Characteristics of Noble-Rot Wines"; R. C. Cooke et al., "Quantification of Several 4-Alkyl Substituted γ-lactones in Australian Wines," *Journal of Agricultural and Food Chemistry* 57, no. 2 (2009): 348–52; J. Langen et al., "Quantitative Analysis of γ- and δ-lactones in Wines Using Gas Chromatography with Selective Tandem Mass Spectrometric Detection," *Rapid Communications in Mass Spectrometry* 27, no. 24 (2013): 2751–59.
46. López et al., "Analysis of the Aroma Intensities of Volatile Compounds"; Langen et al., "Quantitative Analysis of γ- and δ-lactones in Wines Using Gas Chromatography."

47. Wang et al., "Aroma Compounds and Characteristics of Noble-Rot Wines"; Ferreira, Lopez, and Cacho, "Quantitative Determination of the Odorants of Young Red Wines from Different Grape Varieties"; Langen et al., "Quantitative Analysis of γ- and δ-lactones in Wines Using Gas Chromatography"; S. Nakamura et al., "Quantitative Analysis of γ-nonalactone in Wines and Its Threshold Determination," *Journal of Food Science* 53, no. 4 (1988): 1243–44.
48. Langen et al., "Quantitative Analysis of γ- and δ-lactones in Wines Using Gas Chromatography"; O. Vyviurska and I. Špánik, "Assessment of Tokaj Varietal Wines with Comprehensive Two-Dimensional Gas Chromatography Coupled to High Resolution Mass Spectrometry," *Microchemical Journal* 152 (2020): 104385.
49. M. J. Lacey et al., "Methoxypyrazines in Sauvignon Blanc Grapes and Wines," *American Journal of Enology and Viticulture* 42, no. 2 (1991): 103–8; G. Pickering et al., "Determination of Ortho- and Retronasal Detection Thresholds for 2-Isopropyl-3-Methoxypyrazine in Wine," *Journal of Food Science* 72, no. 7 (2007): S468–S472; D. Sidhu et al., "Methoxypyrazine Analysis and Influence of Viticultural and Enological Procedures on Their Levels in Grapes, Musts, and Wines," *Critical Reviews in Food Science and Nutrition* 55, no. 4 (2015): 485–502.
50. Vilanova et al., "Determination of Odorants in Varietal Wines."
51. Ferreira, Lopez, and Cacho, "Quantitative Determination of the Odorants of Young Red Wines"; A. Buettner, "Investigation of Potent Odorants and Afterodor Development in Two Chardonnay Wines Using the Buccal Odor Screening System (BOSS)," *Journal of Agricultural and Food Chemistry* 52, no. 8 (2004): 2339–46; C. Ou et al., "Volatile Compounds and Sensory Attributes of Wine from cv. Merlot (*Vitis vinifera* L.) Grown Under Differential Levels of Water Deficit with or Without a Kaolin-Based, Foliar Reflectant Particle Film," *Journal of Agricultural and Food Chemistry* 58, no. 24 (2010): 12890–898.
52. López et al., "Analysis of the Aroma Intensities of Volatile Compounds"; Wang et al., "Aroma Compounds and Characteristics of Noble-Rot Wines"; Buettner, "Investigation of Potent Odorants and Afterodor Development in Two Chardonnay Wines."
53. Vitalini et al., "The Application of Chitosan and Benzothiadiazole"; Zhang et al., "Comparative Study of Aromatic Compounds in Young Red Wines"; R. Lopez et al., "Identification of Impact Odorants of Young Red Wines Made with Merlot, Cabernet Sauvignon, and Grenache Grape Varieties: A Comparative Study," *Journal of the Science of Food and Agriculture* 79, no. 11 (1999): 1461–67; Bramley, Ouzman, and Boss, "Variation in Vine Vigour, Grape Yield, and Vineyard Soils and Topography"; Wang et al., "Aroma Compounds and Characteristics of Noble-Rot Wines"; Xi et al., "Impact of Cover Crops"; Ou et al., "Volatile Compounds and Sensory Attributes of Wine"; S.-H. Lee et al., "Vine Microclimate and Norisoprenoid Concentration in Cabernet Sauvignon Grapes and Wines," *American Journal of Enology and Viticulture* 58, no. 3 (2007): 291–301; M. C. Qian, Y. Fang, and K. Shellie, "Volatile Composition of Merlot Wine from Different Vine Water Status," *Journal of Agricultural and Food Chemistry* 57, no. 16 (2009): 7459–63.
54. Lopez et al., "Identification of Impact Odorants of Young Red Wines"; Wang et al., "Aroma Compounds and Characteristics of Noble-Rot Wines"; J. Song et al., "Pinot Noir Wine Composition from Different Vine Vigour Zones Classified by Remote Imaging Technology," *Food Chemistry* 153 (2014): 52–59; Y. Kotseridis et al., "Quantitative Determination of β-ionone in Red Wines and Grapes of Bordeaux Using a Stable Isotope Dilution Assay," *Journal of Chromatography A* 848, no. 1–2 (1999): 317–25; J. d. S. Câmara et al., "Varietal Flavour Compounds of Four Grape Varieties Producing Madeira Wines," *Analytica Chimica Acta* 513, no. 1 (2004): 203–7.
55. M. A. Segurel et al., "Contribution of Dimethyl Sulfide to the Aroma of Syrah and Grenache Noir Wines and Estimation of Its Potential in Grapes of These Varieties," *Journal of Agricultural and Food Chemistry* 52, no. 23 (2004): 7084–93; B. Fedrizzi et al., "Aging Effects and Grape Variety Dependence on the Content of Sulfur Volatiles in Wine," *Journal of Agricultural and Food Chemistry* 55, no. 26 (2007): 10880–87; S. K. Park et al., "Incidence of Volatile Sulfur Compounds in California Wines: A Preliminary Survey," *American Journal of Enology and Viticulture* 45, no. 3 (1994): 341–44.

56. B. Yang, P. Schwarz, and R. Horsley, "Factors Involved in the Formation of Two Precursors of Dimethylsulfide During Malting," *Journal of the American Society of Brewing Chemists* 56, no. 3 (1998): 85–92; C. W. Bamforth, "Dimethyl Sulfide—Significance, Origins, and Control," *Journal of the American Society of Brewing Chemists* 72, no. 3 (2014): 165–68.
57. Lopez et al., "Identification of Impact Odorants of Young Red Wines"; Qian, Fang, and Shellie, "Volatile Composition of Merlot Wine from Different Vine Water Status"; M. González-Álvarez et al., "Impact of Phytosanitary Treatments with Fungicides (Cyazofamid, Famoxadone, Mandipropamid, and Valifenalate) on Aroma Compounds of Godello White Wines," *Food Chemistry* 131, no. 3 (2012): 826–36; M. P. Nikfardjam, B. May, and C. Tschiersch, "Analysis of 4-vinylphenol and 4-vinylguaiacol in Wines from the Wurttemberg Region (Germany)," *Mitteilungen Klosterneuburg, Rebe und Wein, Obstbau und Früchteverwertung* 52, no. 2 (2009): 84–89; E. G. García-Carpintero et al., "Volatile and Sensory Characterization of Red Wines from cv. Moravia Agria Minority Grape Variety Cultivated in La Mancha Region Over Five Consecutive Vintages," *Food Research International* 44, no. 5 (2011): 1549–60.
58. Lopez et al., "Identification of Impact Odorants of Young Red Wines"; Qian, Fang, and Shellie, "Volatile Composition of Merlot Wine from Different Vine Water Status"; García-Carpintero et al., "Volatile and Sensory Characterization of Red Wines"; E. Gómez-Plaza et al., "Investigation on the Aroma of Wines from Seven Clones of Monastrell Grapes," *European Food Research and Technology* 209, nos. 3/4 (1999): 257–60.
59. Lopez et al., "Identification of Impact Odorants of Young Red Wines"; Qian, Fang, and Shellie, "Volatile Composition of Merlot Wine from Different Vine Water Status"; García-Carpintero et al., "Volatile and Sensory Characterization of Red Wines"; Gómez-Plaza et al., "Investigation on the Aroma of Wines from Seven Clones of Monastrell Grapes."
60. Lopez et al., "Identification of Impact Odorants of Young Red Wines"; Wang et al., "Aroma Compounds and Characteristics of Noble-Rot Wines"; García-Carpintero et al., "Volatile and Sensory Characterization of Red Wines."
61. Lopez et al., "Identification of Impact Odorants of Young Red Wines"; Wang et al., "Aroma Compounds and Characteristics of Noble-Rot Wines"; García-Carpintero et al., "Volatile and Sensory Characterization of Red Wines."
62. Poisson and Schieberle, "Characterization of the Most Odor-Active Compounds in an American Bourbon Whisky."
63. Lahne, "Aroma Characterization of American Rye Whiskey."
64. H. H. Jeleń, M. Majcher, and A. Szwengiel, "Key Odorants in Peated Malt Whisky and Its Differentiation from Other Whisky Types Using Profiling of Flavor and Volatile Compounds," *LWT—Food Science and Technology* 107 (2019): 56–63.
65. Vitalini et al., "The Application of Chitosan and Benzothiadiazole"; Baumes et al., "Identification and Determination of Volatile Constituents in Wines."
66. Fan et al., "Identification and Quantification of Impact Aroma Compounds."
67. Bettenhausen et al., "Influence of Malt Source on Beer Chemistry, Flavor, and Flavor Stability."
68. López et al., "Analysis of the Aroma Intensities of Volatile Compounds"; Caven-Quantrill and Buglass, "Seasonal Variation of Flavour Content of English Vineyard Grapes."
69. Dong et al., "Characterization of Volatile Aroma Compounds."
70. Flamini, De Luca, and Di Stefano, "Changes in Carbonyl Compounds."
71. Dong et al., "Characterization of Volatile Aroma Compounds."
72. Vitalini et al., "The Application of Chitosan and Benzothiadiazole"; Flamini, De Luca, and Di Stefano, "Changes in Carbonyl Compounds."
73. Dong et al., "Characterization of Volatile Aroma Compounds"; Cramer et al., "Analysis of Volatile Compounds from Various Types of Barley Cultivars."
74. López et al., "Analysis of the Aroma Intensities of Volatile Compounds."
75. Vişan, Dobrinoiu, and Dănăilă-Guidea, "The Agrobiological Study."
76. Dong et al., "Characterization of Volatile Aroma Compounds."

77. Vitalini et al., "The Application of Chitosan and Benzothiadiazole"; Flamini, De Luca, and Di Stefano, "Changes in Carbonyl Compounds."
78. Dong et al., "Characterization of Volatile Aroma Compounds."
79. Zhang et al., "Comparative Study of Aromatic Compounds in Young Red Wines"; Jiang et al., "Comparison of Aroma Compounds in Cabernet Sauvignon and Merlot Wines"; Cheng et al., "Comparison Between Aroma Compounds."
80. Pozo-Bayón et al., "Effect of Vineyard Yield on the Composition of Sparkling Wines"; Aleixandre et al., "Varietal Differentiation of Red Wines in the Valencian Region (Spain)."
81. Dong et al., "Characterization of Volatile Aroma Compounds."
82. J. Jurado et al., "Differentiation of Certified Brands of Origins of Spanish White Wines by HS-SPME-GC and Chemometrics," *Analytical and Bioanalytical Chemistry* 390, no. 3 (2008): 961–70; R. González-Rodríguez et al., "Application of New Fungicides Under Good Agricultural Practices and Their Effects on the Volatile Profile of White Wines," *Food Research International* 44, no. 1 (2011): 397–403; A. Ziółkowska, E. Wąsowicz, and H. H. Jeleń, "Differentiation of Wines According to Grape Variety and Geographical Origin Based on Volatiles Profiling Using SPME-MS and SPME-GC/MS Methods," *Food Chemistry* 213 (2016): 714–20.
83. Vitalini et al., "The Application of Chitosan and Benzothiadiazole"; Vilanova et al., "Determination of Odorants in Varietal Wines from International Grape Cultivars"; Lopez et al., "Identification of Impact Odorants of Young Red Wines."
84. Lopez et al., "Identification of Impact Odorants of Young Red Wines."
85. Vitalini et al., "The Application of Chitosan and Benzothiadiazole"; Jiang et al., "Comparison of Aroma Compounds in Cabernet Sauvignon and Merlot Wines"; Zhang et al., "Comparative Study of Aromatic Compounds in Young Red Wines"; Lopez et al., "Identification of Impact Odorants of Young Red Wines"; Vilanova and C. Martínez, "First Study of Determination of Aromatic Compounds of Red Wine"; Bramley, Ouzman, and Boss, "Variation in Vine Vigour, Grape Yield, and Vineyard Soils and Topography"; Green et al., "Sensory and Chemical Characterisation of Sauvignon Blanc Wine"; Wang et al., "Aroma Compounds and Characteristics of Noble-Rot Wines."
86. Moio and Etievant, "Ethyl Anthranilate, Ethyl Cinnamate, 2, 3-Dihydrocinnamate, and Methyl Anthranilate"; Fang and Qian, "Quantification of Selected Aroma-Active Compounds in Pinot Noir Wines from Different Grape Maturities"; Ferreira, Lopez, and Cacho, "Quantitative Determination of the Odorants of Young Red Wines from Different Grape Varieties"; Antalick et al., "Influence of Grape Composition on Red Wine Ester Profile."
87. Vitalini et al., "The Application of Chitosan and Benzothiadiazole"; Jiang et al., "Comparison of Aroma Compounds in Cabernet Sauvignon and Merlot Wines"; Cheng et al., "Comparison Between Aroma Compounds"; Zhang et al., "Comparative Study of Aromatic Compounds in Young Red Wines."
88. Bramley, Ouzman, and Boss, "Variation in Vine Vigour, Grape Yield, and Vineyard Soils and Topography"; Fang and Qian, "Quantification of Selected Aroma-Active Compounds."
89. Vitalini et al., "The Application of Chitosan and Benzothiadiazole"; Fang and Qian, "Aroma Compounds in Oregon Pinot Noir Wine."
90. Vitalini et al., "The Application of Chitosan and Benzothiadiazole"; Zhang et al., "Comparative Study of Aromatic Compounds in Young Red Wines"; Xi et al., "Impact of Cover Crops."
91. Lopez et al., "Identification of Impact Odorants of Young Red Wines"; Bramley, Ouzman, and Boss, "Variation in Vine Vigour, Grape Yield, and Vineyard Soils and Topography"; Green et al., "Sensory and Chemical Characterisation of Sauvignon Blanc Wine."
92. Lopez et al., "Identification of Impact Odorants of Young Red Wines"; Aleixandre et al., "Varietal Differentiation of Red Wines in the Valencian Region (Spain)"; Alvarez et al., "Geographical Differentiation of White Wines."
93. Lopez et al., "Identification of Impact Odorants of Young Red Wines"; Zhang et al., "Comparative Study of Aromatic Compounds in Young Red Wines"; Falqué, Fernández, and Dubourdieu, "Differentiation of White Wines by Their Aromatic Index."

94. Zhang et al., "Comparative Study of Aromatic Compounds in Young Red Wines"; Bramley, Ouzman, and Boss, "Variation in Vine Vigour, Grape Yield, and Vineyard Soils and Topography."
95. Bettenhausen et al., "Variation in Sensory Attributes and Volatile Compounds."
96. Lopez et al., "Identification of Impact Odorants of Young Red Wines"; Bramley, Ouzman, and Boss, "Variation in Vine Vigour, Grape Yield, and Vineyard Soils and Topography"; Green et al., "Sensory and Chemical Characterisation of Sauvignon Blanc Wine"; Wang et al., "Aroma Compounds and Characteristics of Noble-Rot Wines."
97. Dong et al., "Characterization of Volatile Aroma Compounds."
98. Pérez-Magariño et al., "Multivariate Analysis for the Differentiation of Sparkling Wines"; Martínez-Pinilla et al., "Characterization of Volatile Compounds and Olfactory Profile of Red Minority Varietal Wines from La Rioja"; Vilanova et al., "Volatile Composition and Sensory Properties of North West Spain White Wines"; Slegers et al., "Volatile Compounds from Grape Skin, Juice, and Wine."
99. Wang et al., "Aroma Compounds and Characteristics of Noble-Rot Wines."
100. Vitalini et al., "The Application of Chitosan and Benzothiadiazole"; Xi et al., "Impact of Cover Crops."
101. Dong et al., "Characterization of Volatile Aroma Compounds."
102. Zhang et al., "Comparative Study of Aromatic Compounds in Young Red Wines"; Arozarena, "Multivariate Differentiation of Spanish Red Wines According to Region and Variety"; Bellincontro et al., "Feasibility of an Electronic Nose."
103. Dong et al., "Characterization of Volatile Aroma Compounds."
104. Bramley, Ouzman, and Boss, "Variation in Vine Vigour, Grape Yield, and Vineyard Soils and Topography"; Green et al., "Sensory and Chemical Characterisation of Sauvignon Blanc Wine"; Wang et al., "Aroma Compounds and Characteristics of Noble-Rot Wines."
105. Robinson et al., "Influence of Geographic Origin on the Sensory Characteristics and Wine Composition of *Vitis vinifera*."
106. Guth, "Identification of Character Impact Odorants of Different White Wine Varieties"; Guth, "Quantitation and Sensory Studies of Character Impact Odorants of Different White Wine Varieties."
107. Dong et al., "Characterization of Volatile Aroma Compounds."
108. Guth, "Quantitation and Sensory Studies of Character Impact Odorants of Different White Wine Varieties"; Siebert et al., "Analysis, Potency, and Occurrence of (Z)-6-dodeceno-γ-lactone in White Wine."
109. Cooke et al., "Quantification of Several 4-Alkyl Substituted γ-lactones in Australian Wines"; Langen et al., "Quantitative Analysis of γ- and δ-lactones in Wines Using Gas Chromatography."
110. López et al., "Analysis of the Aroma Intensities of Volatile Compounds"; Langen et al., "Quantitative Analysis of γ- and δ-lactones in Wines Using Gas Chromatography."
111. Langen et al., "Quantitative Analysis of γ- and δ-lactones in Wines Using Gas Chromatography"; Nakamura et al., "Quantitative Analysis of γ-nonalactone in Wines and Its Threshold Determination."
112. Langen et al., "Quantitative Analysis of γ- and δ-lactones in Wines Using Gas Chromatography"; Nakamura et al., "Quantitative Analysis of γ-nonalactone in Wines and Its Threshold Determination"; Vyviurska and Špánik, "Assessment of Tokaj Varietal Wines with Comprehensive Two-Dimensional Gas Chromatography."
113. Lacey et al., "Methoxypyrazines in Sauvignon Blanc Grapes and Wines"; Pickering et al., "Determination of Ortho- and Retronasal Detection Thresholds for 2-Isopropyl-3-Methoxypyrazine in Wine"; Sidhu et al., "Methoxypyrazine Analysis and Influence of Viticultural and Enological Procedures on Their Levels in Grapes, Musts, and Wines."
114. Vilanova et al., "Determination of Odorants in Varietal Wines from International Grape Cultivars (*Vitis vinifera*) Grown in NW Spain."
115. P. Etièvant et al., "Varietal and Geographic Classification of French Red Wines in Terms of Major Acids," *Journal of the Science of Food and Agriculture* 46, no. 4 (1989): 421–38; M. Gil et al.,

"Characterization of the Volatile Fraction of Young Wines from the Denomination of Origin 'Vinos de Madrid' (Spain)," *Analytica Chimica Acta* 563, nos. 1/2 (2006): 145–53.

116. Wang et al., "Aroma Compounds and Characteristics of Noble-Rot Wines"; Buettner, "Investigation of Potent Odorants and Afterodor Development."
117. Ou et al., "Volatile Compounds and Sensory Attributes of Wine"; Lee et al., "Vine Microclimate and Norisoprenoid Concentration in Cabernet Sauvignon Grapes and Wines"; Qian, Fang, and Shellie, "Volatile Composition of Merlot Wine from Different Vine Water Status."
118. Song et al., "Pinot Noir Wine Composition from Different Vine Vigour Zones"; Kotseridis et al., "Quantitative Determination of β-ionone in Red Wines and Grapes of Bordeaux"; Câmara et al., "Varietal Flavour Compounds of Four Grape Varieties Producing Madeira Wines."
119. Segurel et al., "Contribution of Dimethyl Sulfide to the Aroma of Syrah and Grenache Noir Wines"; Fedrizzi et al., "Aging Effects and Grape Variety Dependence on the Content of Sulfur Volatiles in Wine"; Park et al., "Incidence of Volatile Sulfur Compounds in California Wines."
120. Yang, Schwarz, and Horsley, "Factors Involved in the Formation of Two Precursors of Dimethylsulfide During Malting"; Bamforth, "Dimethyl Sulfide."
121. González-Álvarez et al., "Impact of Phytosanitary Treatments with Fungicides"; Nikfardjam, May, and Tschiersch, "Analysis of 4-Vinylphenol and 4-Vinylguaiacol in Wines"; García-Carpintero et al., "Volatile and Sensory Characterization of Red Wines."
122. García-Carpintero et al., "Volatile and Sensory Characterization of Red Wines"; Gómez-Plaza et al., "Investigation on the Aroma of Wines from Seven Clones of Monastrell Grapes"; V. Ferreira et al., "The Chemical Foundations of Wine Aroma—a Role Game Aiming at Wine Quality, Personality, and Varietal Expression," in *Proceedings of the Thirteenth Australian Wine Industry Technical Conference* (Adelaide: Australian Wine Industry Technical Conference, 2007).
123. García-Carpintero et al., "Volatile and Sensory Characterization of Red Wines"; Gómez-Plaza et al., "Investigation on the Aroma of Wines from Seven Clones of Monastrell Grapes."
124. Lopez et al., "Identification of Impact Odorants of Young Red Wines"; Wang et al., "Aroma Compounds and Characteristics of Noble-Rot Wines"; García-Carpintero et al., "Volatile and Sensory Characterization of Red Wines."
125. Qian, Fang, and Shellie, "Volatile Composition of Merlot Wine from Different Vine Water Status"; García-Carpintero et al., "Volatile and Sensory Characterization of Red Wines."

APPENDIX 4

KEY TO THE ROADMAP: SOURCES FOR CHAPTER 17

1. D. Herb, "The Whisky Terroir Project," 2018, https://waterfordwhisky.com/element/the-whisky-terroir-project/.
2. L. Poisson and P. Schieberle, "Characterization of the Most Odor-Active Compounds in an American Bourbon Whisky by Application of the Aroma Extract Dilution Analysis," *Journal of Agricultural and Food Chemistry* 56, no. 14 (2008): 5813–19; J. Lahne, "Aroma Characterization of American Rye Whiskey by Chemical and Sensory Assays," master's thesis, University of Illinois at Urbana-Champaign, 2010, http://hdl.handle.net/2142/16713; H. H. Jeleń, M. Majcher, and A. Szwengiel, "Key Odorants in Peated Malt Whisky and Its Differentiation from Other Whisky Types Using Profiling of Flavor and Volatile Compounds," *LWT—Food Science and Technology* 107 (2019): 56–63; L. Poisson and P. Schieberle, "Characterization of the Key Aroma Compounds in an American Bourbon Whisky by Quantitative Measurements, Aroma Recombination, and Omission Studies," *Journal of Agricultural and Food Chemistry* 56, no. 14 (2008): 5820–26.
3. R. J. Arnold et al., "Assessing the Impact of Corn Variety and Texas Terroir on Flavor and Alcohol Yield in New-Make Bourbon Whiskey," *PloS One* 14, no. 8 (2019).
4. J. L. Aleixandre et al., "Varietal Differentiation of Red Wines in the Valencian Region (Spain)," *Journal of Agricultural and Food Chemistry* 50, no. 4 (2002): 751–55; M. Pozo-Bayón et al., "Effect of Vineyard Yield on the Composition of Sparkling Wines Produced from the Grape Cultivar Parellada," *Food Chemistry* 86, no. 3 (2004): 413–19.
5. B. Jiang and Z. Zhang, "Volatile Compounds of Young Wines from Cabernet Sauvignon, Cabernet Gernischet, and Chardonnay Varieties Grown in the Loess Plateau Region of China," *Molecules* 15, no. 12 (2010): 9184–96; R. González-Rodríguez et al., "Application of New Fungicides Under Good Agricultural Practices and Their Effects on the Volatile Profile of White Wines," *Food Research International* 44, no. 1 (2011): 397–403.
6. J. Bueno et al., "Selection of Volatile Aroma Compounds by Statistical and Enological Criteria for Analytical Differentiation of Musts and Wines of Two Grape Varieties," *Journal of Food Science* 68, no. 1 (2003): 158–63; E. Falqué, E. Fernández, and D. Dubourdieu, "Differentiation of White Wines by Their Aromatic Index," *Talanta* 54, no. 2 (2001): 271–81; S. Pérez-Magariño et al., "Multivariate Analysis for the Differentiation of Sparkling Wines Elaborated from Autochthonous Spanish Grape Varieties: Volatile Compounds, Amino Acids, and Biogenic Amines," *European Food Research and Technology* 236, no. 5 (2013): 827–41.

7. Jiang and Zhang, "Volatile Compounds of Young Wines"; V. Ferreira, R. López, and J. F. Cacho, "Quantitative Determination of the Odorants of Young Red Wines from Different Grape Varieties," *Journal of the Science of Food and Agriculture* 80, no. 11 (2000): 1659–67.
8. Bueno et al., "Selection of Volatile Aroma Compounds"; Falqué, Fernández, and Dubourdieu, "Differentiation of White Wines by Their Aromatic Index"; Pérez-Magariño et al., "Multivariate Analysis for the Differentiation of Sparkling Wines"; N. Moreira et al., "Relationship Between Nitrogen Content in Grapes and Volatiles, Namely Heavy Sulphur Compounds, in Wines," *Food Chemistry* 126, no. 4 (2011): 1599–1607; B. T. Weldegergis, A. de Villiers, and A. M. Crouch, "Chemometric Investigation of the Volatile Content of Young South African Wines," *Food Chemistry* 128, no. 4 (2011): 1100–1109.
9. Bueno et al., "Selection of Volatile Aroma Compounds."

NOTES

INTRODUCTION

1. Whether the word is spelled "whiskey" or "whisky" is based on where the spirit is made and sometimes simply on the preference of the distiller. The plural of "whisky" is "whiskies," and the plural of "whiskey" is "whiskeys." Traditionally, Irish and American distillers use the "e," and everyone else does not. It wasn't until the nineteenth century that Irish and American distillers began to adopt the "e," largely to differentiate themselves from Scottish and Canadian distillers. In this book, unless discussing scotch or Canadian or Japanese whisky specifically, the default spelling will be "whiskey."
2. E. Vaudour, "The Quality of Grapes and Wine in Relation to Geography: Notions of Terroir at Various Scales," *Journal of Wine Research* 13, no. 2 (2002): 117–41.
3. N. Parrott, N. Wilson, and J. Murdoch, "Spatializing Quality: Regional Protection and the Alternative Geography of Food," *European Urban and Regional Studies* 9, no. 3 (2002): 241–61.

2. THE PRODUCTION AND PERCEPTION OF FLAVOR

1. M. Meister, "On the Dimensionality of Odor Space," *Elife* 4 (2015): e07865.
2. P. Polášková, J. Herszage, and S. E. Ebeler, "Wine Flavor: Chemistry in a Glass," *Chemical Society Reviews* 37, no. 11 (2008): 2478–89.
3. D. Tieman, "A Chemical Genetic Roadmap to Improved Tomato Flavor," *Science* 355, no. 6323 (2017): 391–94.

3. THE CHEMISTRY OF FLAVOR

1. J. Goode, "Wine Flavour Chemistry," 2011, https://www.guildsomm.com/public_content/features/articles/b/jamie_goode/posts/wine-flavour-chemistry.
2. V. Ferreira et al., "The Chemical Foundations of Wine Aroma—a Role Game Aiming at Wine Quality, Personality, and Varietal Expression," in *Proceedings of the Thirteenth Australian Wine Industry Technical Conference* (Adelaide: Australian Wine Industry Technical Conference, 2007).

3. THE CHEMISTRY OF FLAVOR

3. Ferreira et al., "The Chemical Foundations of Wine Aroma."
4. Ferreira et al., "The Chemical Foundations of Wine Aroma"; M. C. Goldner et al., "Effect of Ethanol Level in the Perception of Aroma Attributes and the Detection of Volatile Compounds in Red Wine," *Journal of Sensory Studies* 24, no. 2 (2009): 243–57.
5. A. L. Robinson et al., "Origins of Grape and Wine Aroma, Part 1: Chemical Components and Viticultural Impacts," *American Journal of Enology and Viticulture* 65, no. 1 (2014): 1–24.
6. Ferreira et al., "The Chemical Foundations of Wine Aroma."
7. A. D. Webb and R. E. Kepner, "Fusel Oil Analysis by Means of Gas-Liquid Partition Chromatography," *American Journal of Enology and Viticulture* 12, no. 2 (1961): 51–59.
8. L. A. Hazelwood et al., "The Ehrlich Pathway for Fusel Alcohol Production: A Century of Research on *Saccharomyces cerevisiae* Metabolism," *Applied Environmental Microbiology* 74, no. 8 (2008): 2259–66.
9. S. Engan, "Wort Composition and Beer Flavour, Part II: The Influence of Different Carbohydrates on the Formation of Some Flavour Components During Fermentation," *Journal of the Institute of Brewing* 78, no. 2 (1972): 169–73.
10. M. M. Mendes-Pinto, "Carotenoid Breakdown Products: The Norisoprenoids in Wine Aroma," *Archives of Biochemistry and Biophysics* 483, no. 2 (2009): 236–45.
11. Mendes-Pinto, "Carotenoid Breakdown Products."
12. H. H. Jeleń, M. Majcher, and A. Szwengiel, "Key Odorants in Peated Malt Whisky and Its Differentiation from Other Whisky Types Using Profiling of Flavor and Volatile Compounds," *LWT—Food Science and Technology* 107 (2019): 56–63; J. Lahne, "Aroma Characterization of American Rye Whiskey by Chemical and Sensory Assays," master's thesis, University of Illinois at Urbana-Champaign, 2010, http://hdl.handle.net/2142/16713; L. Poisson and P. Schieberle, "Characterization of the Key Aroma Compounds in an American Bourbon Whisky by Quantitative Measurements, Aroma Recombination, and Omission Studies," *Journal of Agricultural and Food Chemistry* 56, no. 14 (2008): 5820–26; L. Poisson and P. Schieberle, "Characterization of the Most Odor-Active Compounds in an American Bourbon Whisky by Application of the Aroma Extract Dilution Analysis," *Journal of Agricultural and Food Chemistry* 56, no. 14 (2008): 5813–19.
13. J. Song et al., "Effect of Grape Bunch Sunlight Exposure and UV Radiation on Phenolics and Volatile Composition of *Vitis vinifera* L. Cv. Pinot Noir Wine," *Food Chemistry* 173 (2015): 424–31.
14. Jeleń, Majcher, and Szwengiel, "Key Odorants in Peated Malt Whisky"; Lahne, "Aroma Characterization of American Rye Whiskey"; Poisson and Schieberle, "Characterization of the Key Aroma Compounds in an American Bourbon Whisky"; Poisson and Schieberle, "Characterization of the Most Odor-Active Compounds in an American Bourbon Whisky."
15. Lahne, "Aroma Characterization of American Rye Whiskey"; Poisson and Schieberle, "Characterization of the Key Aroma Compounds in an American Bourbon Whisky"; Poisson and Schieberle, "Characterization of the Most Odor-Active Compounds in an American Bourbon Whisky."
16. E. G. LaRoe and P. A. Shipley, "Whiskey Composition: Formation of Alpha- and Beta-ionone by the Thermal Decomposition of Beta-carotene," *Journal of Agricultural and Food Chemistry* 18, no. 1 (1970): 174–75.
17. Poisson and Schieberle, "Characterization of the Most Odor-Active Compounds in an American Bourbon Whisky."
18. Ferreira et al., "The Chemical Foundations of Wine Aroma."
19. N. Vanbeneden et al., "Variability in the Release of Free and Bound Hydroxycinnamic Acids from Diverse Malted Barley (*Hordeum vulgare* L.) Cultivars During Wort Production," *Journal of Agricultural and Food Chemistry* 55, no. 26 (2007): 11002–10.
20. B. Boswell, *International Barrel Symposium* (Lebanon, MO: Independent Stave Company, 2008), 1:98–223.
21. J. Gollihue, M. Richmond, H. Wheatley, et al., "Liberation of Recalcitrant Cell Wall Sugars from Oak Barrels Into Bourbon Whiskey During Aging," *Scientific Reports* 8, no. 1 (October 26, 2018): 1–2.

22. B. Harrison et al., "Differentiation of Peats Used in the Preparation of Malt for Scotch Whisky Production Using Fourier Transform Infrared Spectroscopy," *Journal of the Institute of Brewing* 112, no. 4 (2006): 333–39; B. M. Harrison and F. G. Priest, "Composition of Peats Used in the Preparation of Malt for Scotch Whisky Production: Influence of Geographical Source and Extraction Depth," *Journal of Agricultural and Food Chemistry* 57, no. 6 (2009): 2385–391.
23. Poisson and Schieberle, "Characterization of the Key Aroma Compounds in an American Bourbon Whisky"; Poisson and Schieberle, "Characterization of the Most Odor-Active Compounds in an American Bourbon Whisky"; R. J. Arnold et al., "Assessing the Impact of Corn Variety and Texas Terroir on Flavor and Alcohol Yield in New-Make Bourbon Whiskey," *PloS One* 14, no. 8 (2019).
24. V. Ferreira, "Determination of Important Odor-Active Aldehydes of Wine Through Gas Chromatography–Mass Spectrometry of Their O-(2, 3, 4, 5, 6-pentafluorobenzyl) Oximes Formed Directly in the Solid Phase Extraction Cartridge Used for Selective Isolation," *Journal of Chromatography A* 1028, no. 2 (2004): 339–45.

4. THE WINE TERROIR TASTING

1. V. Ferreira et al., "The Chemical Foundations of Wine Aroma—a Role Game Aiming at Wine Quality, Personality, and Varietal Expression," in *Proceedings of the Thirteenth Australian Wine Industry Technical Conference* (Adelaide: Australian Wine Industry Technical Conference, 2007).
2. J. Goode, "Wine Flavour Chemistry," 2011, https://www.guildsomm.com/public_content/features/articles/b/jamie_goode/posts/wine-flavour-chemistry.

6. THE EVOLUTIONARY ROLE OF TERROIR

1. N. P. Hardeman, *Shucks, Shocks, and Hominy Blocks: Corn as a Way of Life in Pioneer America* (Baton Rouge, LA: LSU Press, 1999).
2. I. Groman-Yaroslavski, I. E. Weiss, and D. Nadel, "Composite Sickles and Cereal Harvesting Methods at 23,000-Years-Old Ohalo II, Israel," *PloS One* 11, no. 11 (2016).
3. J. Mercader, "Mozambican Grass Seed Consumption During the Middle Stone Age," *Science* 326, no. 5960 (2009): 1680–83.
4. S. Vimolmangkang et al., "Transcriptome Analysis of the Exocarp of Apple Fruit Identifies Light-Induced Genes Involved in Red Color Pigmentation," *Gene* 534, no. 1 (2014): 78–87.
5. M. E. Kislev, A. Hartmann, and O. Bar-Yosef, "Early Domesticated Fig in the Jordan Valley," *Science* 312, no. 5778 (2006): 1372–74.

7. THE RISE OF COMMODITIES

1. K. Lawson, "The Latest Crop in the Local Food Movement? Wheat," *Modern Farmer*, July 13, 2016, https://modernfarmer.com/2016/07/wheat-terroir/.
2. D. J. Navazio, "Debunking the Hybrid Myth," *Heritage Farm Companion*, 2012, https://www.seedsavers.org/site/pdf/HeritageFarmCompanion_Navazio.pdf.
3. H. J. Klee and D. M. Tieman, "Genetic Challenges of Flavor Improvement in Tomato," *Trends in Genetics* 29, no. 4 (2013): 257–62.

8. A TEXAS TIC-TAC-TOE

1. H. J. Klee, "Improving the Flavor of Fresh Fruits: Genomics, Biochemistry, and Biotechnology," *New Phytologist* 187, no. 1 (2010): 44–56; H. J. Klee and D. M. Tieman, "Genetic Challenges

of Flavor Improvement in Tomato," *Trends in Genetics* 29, no. 4 (2013): 257–62; D. Tieman et al., "A Chemical Genetic Roadmap to Improved Tomato Flavor," *Science* 355, no. 6323 (2017): 391–94; D. Wang and G. B. Seymour, "Tomato Flavor: Lost and Found?," *Molecular Plant* 10, no. 6 (2017): 782–84.
2. A. A. Jaradat and W. Goldstein, "Diversity of Maize Kernels from a Breeding Program for Protein Quality, Part I: Physical, Biochemical, Nutrient, and Color Traits," *Crop Science* 53, no. 3 (2013): 956–76; A. Singh et al., "Nature of the Genetic Variation in an Elite Maize Breeding Cross," *Crop Science* 51, no. 1 (2011): 75–83; B. Shiferaw et al., "Crops That Feed the World, Part 6: Past Successes and Future Challenges to the Role Played by Maize in Global Food Security," *Food Security* 3, no. 3 (2011): 307.
3. J. Doebley et al., "The Origin of Cornbelt Maize: The Isozyme Evidence," *Economic Botany* 42, no. 1 (1988): 120–31; A. F. Troyer, "Background of US Hybrid Corn," *Crop Science* 39, no. 3 (1999): 601–26; A. F. Troyer, "Background of US Hybrid Corn II," *Crop Science* 44, no. 2 (2004): 370–80.
4. M. M. Goodman and W. L. Brown, "Races of Corn," *Corn and Corn Improvement* 18 (1988): 33–79.
5. B. Kurtz et al., "Global Access to Maize Germplasm Provided by the US National Plant Germplasm System and by US Plant Breeders," *Crop Science* 56, no. 3 (2016): 931–41.

9. THE CHEMISTRY OF TERROIR

1. F. Jack, "Development of Guidelines for the Preparation and Handling of Sensory Samples in the Scotch Whisky Industry," *Journal of the Institute of Brewing* 109, no. 2 (2003): 114–19.
2. J. Goode, "Wine Flavour Chemistry," *GuildSomm*, September 7, 2011, https://www.guildsomm.com/public_content/features/articles/b/jamie_goode/posts/wine-flavour-chemistry.

10. THE ROADMAP

1. C. Carlton, "Growing Grains: Bourbon Is Fertilizing a New Market," *Kentucky Living*, March 27, 2018.
2. V. Ferreira et al., "The Chemical Foundations of Wine Aroma—a Role Game Aiming at Wine Quality, Personality, and Varietal Expression," in *Proceedings of the Thirteenth Australian Wine Industry Technical Conference* (Adelaide: Australian Wine Industry Technical Conference, 2007); J. Goode, "Wine Flavour Chemistry," 2011, https://www.guildsomm.com/public_content/features/articles/b/jamie_goode/posts/wine-flavour-chemistry.
3. R. López et al., "Identification of Impact Odorants of Young Red Wines Made with Merlot, Cabernet Sauvignon, and Grenache Grape Varieties: A Comparative Study," *Journal of the Science of Food and Agriculture* 79, no. 11 (1999): 1461–67.
4. Z.-m. Xi et al., "Impact of Cover Crops in Vineyard on the Aroma Compounds of *Vitis vinifera* L. Cv. Cabernet Sauvignon Wine," *Food Chemistry* 127, no. 2 (2011): 516–22.
5. S.-H. Lee et al., "Vine Microclimate and Norisoprenoid Concentration in Cabernet Sauvignon Grapes and Wines," *American Journal of Enology and Viticulture* 58, no. 3 (2007): 291–301.
6. A. Wanikawa, K. Hosoi, and T. Kato, "Conversion of Unsaturated Fatty Acids to Precursors of γ-lactones by Lactic Acid Bacteria During the Production of Malt Whisky," *Journal of the American Society of Brewing Chemists* 58, no. 2 (2000): 51–56; A. Wanikawa et al., "Detection of γ-Lactones in Malt Whisky," *Journal of the Institute of Brewing* 106, no. 1 (2000): 39–44.
7. T. Ilc, D. Werck-Reichhart, and N. Navrot, "Meta-analysis of the Core Aroma Components of Grape and Wine Aroma," *Frontiers in Plant Science* 7 (2016): 1472.
8. X.-j. Wang et al., "Aroma Compounds and Characteristics of Noble-Rot Wines of Chardonnay Grapes Artificially Botrytized in the Vineyard," *Food Chemistry* 226 (2017): 41–50.

9. V. Ferreira, R. López, and J. F. Cacho, "Quantitative Determination of the Odorants of Young Red Wines from Different Grape Varieties," *Journal of the Science of Food and Agriculture* 80, no. 11 (2000): 1659–67.
10. W. Fan et al., "Identification and Quantification of Impact Aroma Compounds in 4 Nonfloral *Vitis vinifera* Varieties Grapes," *Journal of Food Science* 75, no. 1 (2010): S81–S88.
11. L. Dong et al., "Characterization of Volatile Aroma Compounds in Different Brewing Barley Cultivars," *Journal of the Science of Food and Agriculture* 95, no. 5 (2015): 915–21.
12. A.-C. J. Cramer et al., "Analysis of Volatile Compounds from Various Types of Barley Cultivars," *Journal of Agricultural and Food Chemistry* 53, no. 19 (2005): 7526–31.
13. Dong et al., "Characterization of Volatile Aroma Compounds."
14. G. J. Pickering et al., "The Influence of *Harmonia axyridis* on Wine Composition and Aging," *Journal of Food Science* 70, no. 2 (2005): S128–S135.
15. D. Sidhu et al., "Methoxypyrazine Analysis and Influence of Viticultural and Enological Procedures on Their Levels in Grapes, Musts, and Wines," *Critical Reviews in Food Science and Nutrition* 55, no. 4 (2015): 485–502.
16. N. Vanbeneden et al., "Variability in the Release of Free and Bound Hydroxycinnamic Acids from Diverse Malted Barley (*Hordeum vulgare* L.) Cultivars During Wort Production," *Journal of Agricultural and Food Chemistry* 55, no. 26 (2007): 11002–10.
17. Y. He et al., "Wort Composition and Its Impact on the Flavour-Active Higher Alcohol and Ester Formation of Beer—a Review," *Journal of the Institute of Brewing* 120, no. 3 (2014): 157–63.
18. F. Badotti et al., "*Oenococcus alcoholitolerans* sp. nov., a Lactic Acid Bacteria Isolated from Cachaça and Ethanol Fermentation Processes," *Antonie van Leeuwenhoek* 106, no. 6 (2014): 1259–67.
19. K. L. Simpson, B. Pettersson, and F. G. Priest, "Characterization of Lactobacilli from Scotch Malt Whisky Distilleries and Description of *Lactobacillus ferintoshensis* sp. nov., a New Species Isolated from Malt Whisky Fermentations," *Microbiology* 147, no. 4 (2001): 1007–16.
20. P. Costello et al., "Synthesis of Fruity Ethyl Esters by Acyl Coenzyme A: Alcohol Acyltransferase and Reverse Esterase Activities in *Oenococcus oeni* and *Lactobacillus plantarum*," *Journal of Applied Microbiology* 114, no. 3 (2013): 797–806.
21. S. van Beek and F. G. Priest, "Evolution of the Lactic Acid Bacterial Community During Malt Whisky Fermentation: A Polyphasic Study," *Applied Environmental Microbiology* 68, no. 1 (2002): 297–305.
22. Simpson, Pettersson, and Priest, "Characterization of Lactobacilli from Scotch Malt Whisky Distilleries"; van Beek and F. G. Priest, "Evolution of the Lactic Acid Bacterial Community."
23. R. Mitenbuler, *Bourbon Empire: The Past and Future of America's Whiskey* (New York: Penguin, 2016).

11. OVERLAYING THE MAPS

1. J. Lahne, "Aroma Characterization of American Rye Whiskey by Chemical and Sensory Assays," master's thesis, University of Illinois at Urbana-Champaign, 2010, http://hdl.handle.net/2142/16713.
2. M. Lehtonen, "Phenols in Whisky," *Chromatographia* 16, no. 1 (1982): 201–3.
3. C. Scholtes, S. Nizet, and S. Collin, "Guaiacol and 4-Methylphenol as Specific Markers of Torrefied Malts: Fate of Volatile Phenols in Special Beers Through Aging," *Journal of Agricultural and Food Chemistry* 62, no. 39 (2014): 9522–28.
4. F. Vriesekoop et al., "125th Anniversary Review: Bacteria in Brewing: The Good, the Bad, and the Ugly," *Journal of the Institute of Brewing* 118, no. 4 (2012): 335–45.
5. H. H. Jeleń, M. Majcher, and A. Szwengiel, "Key Odorants in Peated Malt Whisky and Its Differentiation from Other Whisky Types Using Profiling of Flavor and Volatile Compounds," *LWT—Food Science and Technology* 107 (2019): 56–63.

11. OVERLAYING THE MAPS

6. B. Harrison et al., "Differentiation of Peats Used in the Preparation of Malt for Scotch Whisky Production Using Fourier Transform Infrared Spectroscopy," *Journal of the Institute of Brewing* 112, no. 4 (2006): 333–39; B. M. Harrison and F. G. Priest, "Composition of Peats Used in the Preparation of Malt for Scotch Whisky Production: Influence of Geographical Source and Extraction Depth," *Journal of Agricultural and Food Chemistry* 57, no. 6 (2009): 2385–91.
7. E. Sihto and V. Arkima, "Proportions of Some Fusel Oil Components in Beer and Their Effect on Aroma," *Journal of the Institute of Brewing* 69, no. 1 (1963): 20–25; C. Van Wyk et al., "Isoamyl Acetate—a Key Fermentation Volatile of Wines of *Vitis vinifera* Cv. Pinotage," *American Journal of Enology and Viticulture* 30, no. 3 (1979): 167–73.
8. V. Ferreira et al., "The Chemical Foundations of Wine Aroma—a Role Game Aiming at Wine Quality, Personality, and Varietal Expression," in *Proceedings of the Thirteenth Australian Wine Industry Technical Conference* (Adelaide: Australian Wine Industry Technical Conference, 2007).
9. Ferreira et al., "The Chemical Foundations of Wine Aroma."
10. D. Langos and M. Granvogl, "Studies on the Simultaneous Formation of Aroma-Active and Toxicologically Relevant Vinyl Aromatics from Free Phenolic Acids During Wheat Beer Brewing," *Journal of Agricultural and Food Chemistry* 64, no. 11 (2016): 2325–32; K. J. Schwarz, L. I. Boitz, and F. J. Methner, "Enzymatic Formation of Styrene During Wheat Beer Fermentation Is Dependent on Pitching Rate and Cinnamic Acid Content," *Journal of the Institute of Brewing* 118, no. 3 (2012): 280–84.
11. V. A. Watts and C. E. Butzke, "Analysis of Microvolatiles in Brandy: Relationship Between Methylketone Concentration and Cognac Age," *Journal of the Science of Food and Agriculture* 83, no. 11 (2003): 1143–49.
12. V. Ferreira et al., "Chemical Characterization of the Aroma of Grenache Rose Wines: Aroma Extract Dilution Analysis, Quantitative Determination, and Sensory Reconstitution Studies," *Journal of Agricultural and Food Chemistry* 50, no. 14 (2002): 4048–54.
13. Lahne, "Aroma Characterization of American Rye Whiskey."
14. L. Poisson and P. Schieberle, "Characterization of the Key Aroma Compounds in an American Bourbon Whisky by Quantitative Measurements, Aroma Recombination, and Omission Studies," *Journal of Agricultural and Food Chemistry* 56, no. 14 (2008): 5820–26.

12. WHISKEY IN THE BIG APPLE

1. W. Curtis, "How Rye Came Back," *The Atlantic*, September 2014.
2. K. Willcox, "New York Farmers and Distillers Reinventing Whiskey," *Edible Capital District*, January 8, 2019, https://ediblecapitaldistrict.ediblecommunities.com/drink/new-york-farmers-and-distillers-empire-rye-whiskey.
3. Available from http://kingscountydistillery.com/about/.
4. P. Hobbs, "Field to Fork: Corn, from the Stalk at Lakeview Organic Grain, to Spirit, at Kings County Distillery," *Nona Brooklyn*, February 13, 2013, http://nonabrooklyn.com/field-to-fork-corn-from-the-stalk-at-lakeview-organic-grains-to-spirit-at-kings-county-distillery/.
5. Hobbs, "Field to Fork."
6. Hobbs, "Field to Fork."
7. Hobbs, "Field to Fork."

13. THE TRILOGY OF FARMING

1. J. Cox, "Organic Tastings: A Great Wine Is One That Gives Great Pleasure," *Rodale's Organic Life*, December 22, 2010, http://www.rodalesorganiclife.com/food/organic-wine.
2. A. Morganstern, "Biodynamics in the Vineyard," *Organic Wine Journal*, March 2008.

3. M. A. Delmas, O. Gergaud, and J. Lim, "Does Organic Wine Taste Better? An Analysis of Experts' Ratings," *Journal of Wine Economics* 11, no. 3 (2016): 329–54.
4. I. Zarraonaindia et al., "The Soil Microbiome Influences Grapevine-Associated Microbiota," *MBio* 6, no. 2 (2015): e02527-14; N. A. Bokulich et al., "Microbial Biogeography of Wine Grapes Is Conditioned by Cultivar, Vintage, and Climate," *Proceedings of the National Academy of Sciences* 111, no. 1 (2014): E139–E148.
5. J. R. Reeve et al., "Soil and Winegrape Quality in Biodynamically and Organically Managed Vineyards," *American Journal of Enology and Viticulture* 56, no. 4 (2005): 367–76.
6. J. K. Reilly, "Moonshine, Part 2: A Blind Sampling of 20 Wines Shows That Biodynamics Works. But How?," *Fortune*, August 23, 2004, 1–2.
7. Delmas, Gergaud, and Lim, "Does Organic Wine Taste Better?"
8. C. F. Ross et al., "Difference Testing of Merlot Produced from Biodynamically and Organically Grown Wine Grapes," *Journal of Wine Research* 20, no. 2 (2009): 85–94.
9. J. Döring et al., "Organic and Biodynamic Viticulture Affect Biodiversity and Properties of Vine and Wine: A Systematic Quantitative Review," *American Journal of Enology and Viticulture* 70, no. 3 (2019): 221–42.

14. MY OLD KENTUCKY HOME

1. C. Carlton, "Growing Grains: Bourbon Is Fertilizing New Market," *Kentucky Living*, March 27, 2018, https://www.kentuckyliving.com/news/growing-grains.
2. KyCorn Growers Association, https://www.kycorn.org/distilled-spirits#:~:text=Bourbon%20%26%20distilled%20Spirits,back%20more%20than%20200%20years.
3. USDA, "Kentucky Corn Production May Set a Record," press release, August 12, 2019, https://www.nass.usda.gov/Statistics_by_State/Kentucky/Publications/Current_News_Release/2019/PRAUG19_KY.pdf.
4. C. Cowdery, "Jim Beam Is Filling 500,000 Barrels a Year. That Is the Real Story," *Chuck Cowdery Blog*, May 2, 2016, http://chuckcowdery.blogspot.com/2016/05/jim-beam-is-filling-500000-barrels-year.html.
5. "Gallo Company Fact Sheet," http://www.gallo.com/files/Gallo-Company-Fact-Sheet-2017.pdf.
6. C. Cowdery, "You Call Yourself 'Craft'? Make Your Own Yeast," *Chuck Cowdery Blog*, February 6, 2011, https://chuckcowdery.blogspot.com/2011/02/you-call-yourself-craft-make-your-own.html.

15. CORN, WHEAT, AND RYE AMONG THE BLUEGRASS

1. S. A. Goff and H. J. Klee, "Plant Volatile Compounds: Sensory Cues for Health and Nutritional Value?," *Science* 311, no. 5762 (2006): 815–19.
2. H. Alem et al., "Impact of Agronomic Practices on Grape Aroma Composition: A Review," *Journal of the Science of Food and Agriculture* 99, no. 3 (2019): 975–85.
3. Alem et al., "Impact of Agronomic Practices on Grape Aroma Composition."
4. Alem et al., "Impact of Agronomic Practices on Grape Aroma Composition."
5. Alem et al., "Impact of Agronomic Practices on Grape Aroma Composition."

16. ACROSS THE POND AND THROUGH THE HILLS

1. D. Herb et al., "Effects of Barley (*Hordeum vulgare* L.) Variety and Growing Environment on Beer Flavor," *Journal of the American Society of Brewing Chemists* 75, no. 4 (2017): 345–53.

16. ACROSS THE POND AND THROUGH THE HILLS

2. P. Hayes, "Barley No Longer an Afterthought in Beer Flavor," Oregon State University Extension Service, November 2017, https://extension.oregonstate.edu/news/barley-no-longer-afterthought-beer-flavor.
3. D. Herb et al., "Malt Modification and Its Effects on the Contributions of Barley Genotype to Beer Flavor," *Journal of the American Society of Brewing Chemists* 75, no. 4 (2017): 354–62.
4. E. G. LaRoe and P. A. Shipley, "Whiskey Composition: Formation of Alpha- and Beta-ionone by the Thermal Decomposition of Beta-carotene," *Journal of Agricultural and Food Chemistry* 18, no. 1 (1970): 174–75.

17. TĒIREOIR

1. H. H. Jeleń, M. Majcher, and A. Szwengiel, "Key Odorants in Peated Malt Whisky and Its Differentiation from Other Whisky Types Using Profiling of Flavor and Volatile Compounds," *LWT—Food Science and Technology* 107 (2019): 56–63.
2. K. L. Simpson, B. Pettersson, and F. G. Priest, "Characterization of Lactobacilli from Scotch Malt Whisky Distilleries and Description of *Lactobacillus ferintoshensis* sp. nov., a New Species Isolated from Malt Whisky Fermentations," *Microbiology* 147, no. 4 (2001): 1007–16.
3. L. Poisson and P. Schieberle, "Characterization of the Most Odor-Active Compounds in an American Bourbon Whisky by Application of the Aroma Extract Dilution Analysis," *Journal of Agricultural and Food Chemistry* 56, no. 14 (2008): 5813–19; J. Lahne, "Aroma Characterization of American Rye Whiskey by Chemical and Sensory Assays," master's thesis, University of Illinois at Urbana-Champaign, 2010, http://hdl.handle.net/2142/16713; Jeleń, Majcher, and Szwengiel, "Key Odorants in Peated Malt Whisky"; L. Poisson and P. Schieberle, "Characterization of the Key Aroma Compounds in an American Bourbon Whisky by Quantitative Measurements, Aroma Recombination, and Omission Studies," *Journal of Agricultural and Food Chemistry* 56, no. 14 (2008): 5820–26.
4. Poisson and Schieberle, "Characterization of the Most Odor-Active Compounds in an American Bourbon Whisky"; Lahne, "Aroma Characterization of American Rye Whiskey."

18. CULTIVATING FLAVOR ON THE FARMS OF ÉIRE

1. S. Duggan, "Seamus Duggan, Durrow, Co. Laois," Waterford Distillery, September 17, 2019, YouTube video, https://www.youtube.com/watch?v=bbqUPbkt9cU2019, YouTube video.
2. Waterford Distillery, "The Jacksons, Organic Barley Growers, Co. Tipperary," May 18, 2019, YouTube video, https://www.youtube.com/watch?v=1soSWe7QryU.

19. AT LAST, A SIP

1. M. Bylok, "Mark Reynier Is Doubling Down on Whisky Terroir," *Whisky Buzz* blog, September 4, 2015, https://whisky.buzz/blog/mark-reynier-is-doubling-down-on-whisky-terroir.

CONCLUSION

1. J. Carswell, "Growing Barley at Coull Farm," 2016, https://www.bruichladdich.com/bruichladdich-whisky-news/barley/growing-barley-coull-farm/.

INDEX

ARTS AND TRADITIONS OF THE TABLE: PERSPECTIVES ON CULINARY HISTORY

Albert Sonnenfeld, Series Editor

Salt: Grain of Life, Pierre Laszlo, translated by Mary Beth Mader
Culture of the Fork, Giovanni Rebora, translated by Albert Sonnenfeld
French Gastronomy: The History and Geography of a Passion, Jean-Robert Pitte, translated by Jody Gladding
Pasta: The Story of a Universal Food, Silvano Serventi and Françoise Sabban, translated by Antony Shugar
Slow Food: The Case for Taste, Carlo Petrini, translated by William McCuaig
Italian Cuisine: A Cultural History, Alberto Capatti and Massimo Montanari, translated by Áine O'Healy
British Food: An Extraordinary Thousand Years of History, Colin Spencer
A Revolution in Eating: How the Quest for Food Shaped America, James E. McWilliams
Sacred Cow, Mad Cow: A History of Food Fears, Madeleine Ferrières, translated by Jody Gladding
Molecular Gastronomy: Exploring the Science of Flavor, Hervé This, translated by M. B. DeBevoise
Food Is Culture, Massimo Montanari, translated by Albert Sonnenfeld
Kitchen Mysteries: Revealing the Science of Cooking, Hervé This, translated by Jody Gladding
Hog and Hominy: Soul Food from Africa to America, Frederick Douglass Opie
Gastropolis: Food and New York City, edited by Annie Hauck-Lawson and Jonathan Deutsch
Building a Meal: From Molecular Gastronomy to Culinary Constructivism, Hervé This, translated by M. B. DeBevoise
Eating History: Thirty Turning Points in the Making of American Cuisine, Andrew F. Smith
The Science of the Oven, Hervé This, translated by Jody Gladding
Pomodoro! A History of the Tomato in Italy, David Gentilcore
Cheese, Pears, and History in a Proverb, Massimo Montanari, translated by Beth Archer Brombert
Food and Faith in Christian Culture, edited by Ken Albala and Trudy Eden
The Kitchen as Laboratory: Reflections on the Science of Food and Cooking, edited by César Vega, Job Ubbink, and Erik van der Linden
Creamy and Crunchy: An Informal History of Peanut Butter, the All-American Food, Jon Krampner

Let the Meatballs Rest: And Other Stories About Food and Culture, Massimo Montanari, translated by Beth Archer Brombert
The Secret Financial Life of Food: From Commodities Markets to Supermarkets, Kara Newman
Drinking History: Fifteen Turning Points in the Making of American Beverages, Andrew F. Smith
Italian Identity in the Kitchen, or Food and the Nation, Massimo Montanari, translated by Beth Archer Brombert
Fashioning Appetite: Restaurants and the Making of Modern Identity, Joanne Finkelstein
The Land of the Five Flavors: A Cultural History of Chinese Cuisine, Thomas O. Höllmann, translated by Karen Margolis
The Insect Cookbook: Food for a Sustainable Planet, Arnold van Huis, Henk van Gurp, and Marcel Dicke, translated by Françoise Takken-Kaminker and Diane Blumenfeld-Schaap
Religion, Food, and Eating in North America, edited by Benjamin E. Zeller, Marie W. Dallam, Reid L. Neilson, and Nora L. Rubel
Umami: Unlocking the Secrets of the Fifth Taste, Ole G. Mouritsen and Klavs Styrbæk, translated by Mariela Johansen and designed by Jonas Drotner Mouritsen
The Winemaker's Hand: Conversations on Talent, Technique, and Terroir, Natalie Berkowitz
Chop Suey, USA: The Story of Chinese Food in America, Yong Chen
Note-by-Note Cooking: The Future of Food, Hervé This, translated by M. B. DeBevoise
Medieval Flavors: Food, Cooking, and the Table, Massimo Montanari, translated by Beth Archer Brombert
Another Person's Poison: A History of Food Allergy, Matthew Smith
Taste as Experience: The Philosophy and Aesthetics of Food, Nicola Perullo
Kosher USA: How Coke Became Kosher and Other Tales of Modern Food, Roger Horowitz
Chow Chop Suey: Food and the Chinese American Journey, Anne Mendelson
Mouthfeel: How Texture Makes Taste, Ole G. Mouritsen and Klavs Styrbæk, translated by Mariela Johansen
Garden Variety: The American Tomato from Corporate to Heirloom, John Hoenig
Cook, Taste, Learn: How the Evolution of Science Transformed the Art of Cooking, Guy Crosby
Meals Matter: A Radical Economics Through Gastronomy, Michael Symons
The Chile Pepper in China: A Cultural Biography, Brian R. Dott